Translating Technology in Africa
Volume 2: Technicisation

Translating Technology in Africa
Volume 2: Technicisation

Edited by

Richard Rottenburg
Eva Riedke

BRILL

LEIDEN | BOSTON

The Library of Congress Cataloging-in-Publication Data is available online at https://catalog.loc.gov

Typeface for the Latin, Greek, and Cyrillic scripts: "Brill". See and download: brill.com/brill-typeface.

ISBN 978-90-04-68827-8 (paperback)
ISBN 978-90-04-68828-5 (e-book)
DOI:10.1163/9789004688285

Contents

Acknowledgements

The series "Translating Technology in Africa" is largely based on the collaborative research programme "Adaptation and Creativity in Africa" (SPP 1448) funded by the German Research Foundation (DFG). The series was developed at the Wits Institute for Social and Economic Research (WiSER) at the University of the Witwatersrand, Johannesburg, South Africa. It is one of the outputs of WiSER's "STS Africa" research focus.

Figures

Notes on Contributions

Sarah Biecker

PhD, is a social scientist working at the intersection of international relations, political sociology and anthropology. She has conducted fieldwork in Rwanda, Uganda and Germany. Her work explores practices of power and violence, policing, bureaucracy and policy-making. She wrote her doctoral thesis on everyday police practice in Uganda and is co-author of *The Political Anthropology of Internationalised Politics* (2021 Rowman and Littelfield).

Marc Boeckler

is Professor of Economic Geography at Goethe University Frankfurt. His work focuses on geographies of marketisation, the performativity of economics, and the infrastructures and spaces of global circulations. His publications on Africa include *Spatial Practice: Territory, Border and Infrastructure in Africa* (2018 Brill).

Jude Kagoro

PhD, is a social scientist at the Institute for Intercultural and International Studies (InIIS), Bremen University. His current work focuses on Somalia within a larger project on *Knowledge Production in Peace and Security Policy* funded by the German Ministry of Education and Research (BMBF). Previously, he worked on two DFG funded research projects, *Policing in Africa* and *Figurations of Internationalised Rule in Africa.* His published monographs include *Inside an African Police Force: the Uganda Police Examined* (2022 Springer) and *Militarisation in Post-1986 Uganda* (2015 Lit). He has been involved in knowledge transfer initiatives with both the Rwandan and the Ugandan Police.

Jochen Monstadt

is Professor of Governance of Urban Dynamics and Transitions and holds the Chair of Spatial Planning at the Department of Human Geography and Spatial Planning at Utrecht University. His research and teaching interests revolve around the contingent and place-based transformation patterns of cities and how these are mediated by technical infrastructures (energy, water, wastewater, solid waste, transportation, and ICTs). His specific interest lies in how the sociotechnical design and governance of those critical systems shape the sustainability of cities in the global North and South.

Sung-Joon Park

PhD, leads the research group *Medical Anthropology* at the Bernhard-Nocht-Institute for Tropical Medicine in Hamburg, Germany. He has pursued extensive ethnographic research in Sierra Leone, Uganda, and the Democratic Republic of the Congo, concerned with Ebola epidemics, the Covid-19 pandemic, the political ecology of disease outbreaks, the ethics of care, mobility, and trust and mistrust. His work has been published in *Social Science and Medicine*, MAT, JRAI, and *Critical Public Health*. His current research in planetary health is interested in how pathogens behave in damaged ecologies.

Eva Riedke

PhD, is Assistant Professor of Social and Cultural Anthropology at the University of Konstanz. Based on fieldwork in South Africa, East Africa, and Germany, her work has been concerned with tracing post-apartheid controversies and their publics; the coming into being of off-grid infrastructures; their entanglement with processes of financialisation; and with questions of ethics and values in energy transitions. She is currently pursuing a research project that follows solar products—from design and marketing by philanthropically motivated entrepreneurs to distribution and use by those living off or under the grid in rural Kenya. She has contributed a series of publications on the topic.

Richard Rottenburg

is Professor of Science and Technology Studies at the Wits Institute of Social and Economic Research (WiSER), University of the Witwatersrand, Johannesburg. His work has been inspired by renditions of post-foundational social theory and material-semiotic praxiography. The main objects of his inquiries are assemblages of evidence-making (experiments, quantifications), their dependency on knowledge infrastructures, and their entanglement with narrative forms of sense-making and technopolitics. The driving questions are: How do these assemblages shape future-making and the attribution of responsibilities? How do they circulate around the globe? Rottenburg is best known for his 2009 book *Far*-fetched Facts with MIT Press.

Klaus Schlichte

is Professor of International Relations at the University of Bremen. His focus is on global political sociology. He carried out research in Senegal, Mali, Serbia, France, and Uganda, and taught at Science-Po, Paris, the University of Washington, Seattle, and several German universities. He is the author of *In the Shadow of Violence: The Politics of Armed Groups* (2009, Chicago University Press), and the co-editor of *The Historicity of International Politics. Imperialism and the Presence of the Past* (2023, Cambridge University Press) and *Extended Experience: The Political Anthropology of Internationalised rule* (2021 Rowman and Littelfield).

Jannik Schritt

PhD, is a sociocultural anthropologist at the University of Göttingen. He has carried out extensive research at the interface of political anthropology, economic anthropology, as well as science and technology studies, with a regional focus on Niger. His recent publications include a monograph, an edited collection, and a series of articles on Niger, including *Oil-Age Africa: Critical Reflections on Oil Politics, Resource Economies and Extractive Communities* (Brill, 2023) and *Crude Moves: Political Power in Oil-Age Niger* (2024 Göttingen University Press).

Alena Thiel

PhD, is a social anthropologist at IT University Copenhagen with over a decade of ethnographic research experience in Ghana. Her recent work covers Ghana's public sector digitalisation, including innovations in biometric identification technologies, digital census methods, digital addressing, and health information systems. She is co-editor of the edited volume *The Social Life of Health Data* (2024 Palgrave).

Christiane Tristl

PhD, is a heterodox economic geographer at Münster University. Her research focuses on big tech-companies and digital technologies, the marketisation of water in a South-North perspective, critical agri-food-studies, and the unearthing of alternative presents. Her monograph *Turning Water into a Commodity: Digital Innovation and the Private Sector as Development Agent* (forthcoming 2025 Bristol University Press) extends the conjuncture on the topics of private sector dominance in development cooperation in the SDG era, water supply, and digital technologies.

Jonas van der Straeten

is Assistant Professor at the Technology, Innovation and Society Group at Eindhoven University of Technology. He is a trained historian of technology. His research focuses on material practices that are commonly labelled as *informal* and includes studies on electricity provision in East Africa, on self-help construction in Central Asia, and the electrification of three-wheeled transport in South Asia. His most recent publications include a co-edited special issue of the journal *Central Asian Studies* on *Technology, Temporality, and the Study of Central Asia*.

Translating Technology in Africa

Richard Rottenburg

1 Probing a Problematisation: Technology's Circulation and Transmorphism

Starting from the premise that problems are not found but constructed, this introduction asks a first question: what (and how) does a series on *Translating Technology in Africa* problematise? Together, the volumes in the series offer exercises in problematising the role of science and technology in Africa. To do so, they probe the multiple ways in which science and technology are developed, circulated, translated, and entangled with other local and translocal considerations and influences. The volumes show how the development and circulation of technoscience are mutually constitutive, multidirectional, and take place between temporalities (past to present), geographical spaces (urban to rural, country to country) and, most importantly, social spheres. The entanglement of the technoscientific with the political, economic, legal, social and cultural spheres is central to all the chapters. The collected case studies describe concrete empirical events and situations in different places on the African continent. Most of them show variations and combinations of a few different types of practices, such as technoscientific solutions to practical problems, experimental innovations, melioristic interventions, creative adaptations, improvisations, and tweaks. Central to these practices is the act of translation.

This problematisation is likely to elicit some concerns. The practices mentioned above—as they pertain to science and technology—also played a central role in preparing for and supporting European imperialism and colonialism. Today they continue to be implicated in the more hidden ways in which postcolonial forms perpetuate and increase intolerable inequalities, as well as in the destruction of the planet's ecology. This book series proposes a way of addressing exactly these troubling aspects by identifying an approach that avoids binary juxtapositions of "us" versus "them," "local" versus "imported," "familiar" versus "alien", or "scientific" versus "false". Importantly, the series argues against essentialising and ontologising these dualisms.

This willingness to eschew essentialising binaries follows a tradition that can be traced back to the first President of the Republic of Senegal, Leopold Sédar Senghor. In the first decades after the independence of most African countries in the 1960s, the idea that "Europe underdeveloped Africa"—a phrase coined by Walter Rodney in 1972—implied a quest for development on the continent. In those years, the aspiration seemed undeniable and focused on technoscientific development. In January 1974 Senghor expressed this aspiration programmatically at the "Conference of the Ministers of African Member States Responsible for the Application of Science and Technology to Development", organised by UNESCO in Dakar (UNESCO 1974). The pressing issue of the time was to compensate for what colonial governments failed to do, and partly even suppressed, in their mission to establish only those technological infrastructures that were necessary for the extraction of mineral resources and agricultural products. The core idea of Senghor's position was reiterated later by Souleymane Bachir Diagne (2013) and Achille Mbembe (2021). According to the latter, the emergence of the colony as a social form gave rise to zones of creolisation that fostered exchange, dialogue, new imaginaries and opportunities for an alternative version of global modernity for the planet. In Mbembe's rephrasing of one of Frantz Fanon's arguments, it was the establishment of self-ownership through disalienation that would allow for the creation of "a new species of man [sic] and new forms of life" (Mbembe 2021, 55). In this sense, our series on *translating* technology is interested in changes and shifts in the emancipatory capacities associated with technosciences in postcolonial African contexts.

A crucial aspect of these emerging shifts concerns the position of African scientists and engineers within their global networks of colleagues, and the position of their institutions and companies in global exchange and competition. There are significant differences in research facilities and resources between countries on the African continent, and huge disparities between African countries and the most economically affluent countries of the transatlantic region and parts of Asia (Russia, China and Japan). In joint research projects with wealthier research institutions from these countries, African partners often have little say in how a project is designed and what the research questions are. African researchers are then often perceived as collectors of data that is to be subsequently analysed elsewhere. The fact that there is a push for internationalisation, collaboration, and diversity among research funders and institutions in Europe, the US and Japan—demanding that grant applications for research in Africa include researchers from the continent—often does not change the pattern whereby data comes from the South and theories from the North. In a seminal article exploring this persistent postcolonial pattern, Paulin Hountondji (1990) coined the term "extraverted research" for the kind of collaborations where research design, data processing, and theoretical interpretation take place elsewhere—that is where most of the money comes from. This is more than an ethical problem. It is a serious loss for scientific knowledge in general, as different and often more appropriate questions come to the fore when asked from the many and varied African contexts. At the same time, this accurate criticism can easily be trivialised into a problematic assumption: that a line can be drawn between insights that are considered better because they are endogenous and those that are considered less good because they are exogenous. Such an assumption ultimately undermines the idea that we gain new and better insights primarily through the exchange of different perspectives and positions—not by closing them off.

To explore how such exchanges unfold and can be improved, the series *Translating Technology in Africa* prioritises two contextualisations: circulation (as opposed to compartmentalisation) and future orientation (as opposed to inheritance). In other words, the starting point is not a search for an original cause, but a curiosity about the contemporary making of the future. The question then becomes how technoscientific knowledge production can be transformed to better serve a just and sustainable future for all people and the planet, rather than serving the reproduction of capital and a small elite. In short, how might technoscientific knowledge support the ongoing search for the commons (Stengers 2015)?

Although it had a different purpose, there is an earlier version of this question that helps to understand and pursue it in a new direction. The academic field that emerged in Britain in the 1970s and later came to be known as Science and Technology Studies (STS) opened up the possibility of studying science and technology as any other social practice. For its first two decades, STS focused on the highly technicised and industrialised transatlantic countries, only slowly realising that its objects of study spanned the globe. More than twenty years after STS gained prominence and became institutionalised in some of these countries—albeit on a very small scale to date—its problematisations and approaches have been taken up in Latin America, Southeast Asia, and Africa, where they have been critically translated into postcolonial contexts and thus modified to deal with the specificities of these situations (Verran 2002; Warwick 2002; Harding 2011). However, another twenty years later the level of institutionalisation and popularity of the field is noticeably lowest in African universities and research institutions. Simply put, in some parts of the academic campus there is a robust, officially celebrated, and comparatively well-funded drive to advance science, medicine, engineering, mathematics, and computing. The aim here is to do the same science as elsewhere. This drive is largely unaffected by debates in other parts of the campus, namely in the humanities and social sciences departments, where a postcolonial critique of modernist and universalist understandings of technoscience is flourishing. The aim here is to find alternative epistemologies. This quest, in turn, remains largely unresponsive to the deep convictions of colleagues in science and engineering that science is international. One may argue that a similar configuration can be found in most universities around the world. STS provides analytical resources to disrupt this conspicuous configuration by exploring its contradictions and examining the entanglement of the technoscientific sphere with the political, economic, legal, social, and cultural spheres. It does not reduce technoscience to power and money, nor power and money to technoscience. Rather, its objects of study lie somewhere in the middle, in the details of the interstitial practices between scientific universalism, situated knowledge, and critique.

Another noteworthy contribution of STS has been to disturb an influential and seemingly stable understanding of science according to which what matters is "high theory" and "breakthrough experiments" done by a few chosen men of genius. Instead, what really matters are the myriad concrete practices of scientific activity and technological development, the material infrastructures on which they rely, and the complex and temporary assemblages they form. These assemblages have a retroactive effect on the kind of science and technology that is being produced, and are always under the influence of a particular zeitgeist. This is not to say that there is no need to continue high-level theoretical and conceptual work aimed at improving predictions— which are necessarily open to correction (Chalmers 1976)—mainly through experiments (including with new research instrumentation), measurements, and statistical calculations (including via digital technology). But there is more. Framing narratives, as they concern progress and apocalypse, or equality and justice, also play an important role in orientating technoscience, together with the unpredictable intricacies of situational practices. Contingency and serendipity often lead to important turns in the production of technoscience.

While narratives are publicly invoked and shape atmospheres, the importance of unpredictable intricacies is mostly kept invisible. The problem with this invisibility is that it distorts public understanding of what scientific knowledge means. The systematic concealment in public discourse of the contingency of the research process and of the probabilistic nature of any scientific findings leads to a false sense of certainty, and a demand for eternal facts. This political invocation ultimately feeds either a naive faith in science or a hazardous disbelief. Both, the grand narratives of progress and of apocalypse, seem to require either scientific certainty, or its radical opposite, to ensure their public plausibility. But science operates with the assumption that most things remain uncertain and often turn out rather differently than predicted. Science therefore rarely claims full certainty once and for ever but mostly establishes probabilities and continues to seek corrections. If we take into account the inextricable entanglement of the technoscientific with political, economic, legal, social, and cultural spheres, the picture becomes even more complex.

The authors in this series aim to critically examine technoscientific practices thus defined, keeping in mind their contextualisation and focusing on the making of the future rather than the origins. They offer detailed descriptions of technologies at work and engage with STS-inspired concepts and methodologies. They do so without assuming that STS is a canonised body of superior scholarship imported from elsewhere and waiting to be deployed in African contexts. Nor do they assume that there is another predetermined or superior way of knowing how technoscience and politics relate to each other. On the contrary, based on their empirical studies, the authors seek to critically and openly unpack and develop existing scholarship in order to enrich both Africa-related STS and STS in general.

This is important because contemporary governments around the world, including in African countries, are now focused on regulating various technoscientific developments such as artificial intelligence (AI), biometric identification, genome editing, pharmaceuticals, geoengineering and so on. This is understandable: these technoscientific developments promise solutions for, and threaten to pose new challenges to, the core concerns of government—equality, justice, freedom, health, liveable environments, education, and employment. But because these concerns of government are critically linked to socio-economic inequalities and the socio-ecological impacts of capitalism—exploitation, poverty, pollution, climate change and biodiversity loss—the other main focus of government is to regulate the economy. The important point for the argument here is that the regulation of the economy is, in turn, something that directly or indirectly involves technoscience. A good understanding of how technoscience works in context is therefore crucial for democratic participation, government, and corporate decision making. Without a sufficient level of lay expertise, democracy is at risk of falling prey to flawed visions of how science and technology shape and are shaped by the world.

It is equally important that technoscientific experts have an adequate understanding of the presuppositions within which their own practices make sense, of the broader assemblages which their practices and material objects are unavoidably a part of, and last but not least, of the popular understandings of what it is that they actually do. The series aims to trigger a vivid curiosity towards science and technology as a field of social inquiry. It intends to demonstrate that the different settings in which technoscientific solutions are developed and deployed require specific approaches to examine these solutions. The objective, thus, is not to import STS to Africa but to provide some inspiration for how to rewrite STS in and from Africa—inspiration not only to students of the humanities, social sciences, and law, but also to students of the sciences, engineering, and medicine.

None of these goals can be achieved without the active involvement of technoscience. Modernist technoscience as we know it today emerged in Europe in the seventeenth and eighteenth centuries by translating diverse forms of knowledge from around the world into a new assemblage. In this translation process a lot was lost and sacrificed for the sake of epistemological coherence and technological connectivity. During the subsequent two centuries, this assemblage expanded across the entire world under the banner of progress and civilisation—a banner that regularly served as poor camouflage for crimes against humanity, exploitation, and suppression of other ways of knowing.

Modernist technoscience did not prevent, but rather reinforced capitalist mechanisms of self-devouring growth and ruination through reckless industrial production, leading not only to widening inequalities, but also to the accumulation of toxic residues that now threaten human life on the planet. The technosciences of the twenty-first century must create new forms of sustainable future-making that will be measured by their ability to contribute to the healing of the planet. To achieve this goal, local vernacular methodologies and technoscientific archives must be brought together in the making of a world where humanity is better equipped to respond to the catastrophic times in which we live.

2   Probing Concepts: Assemblage, Translation, Transmorphism
       and Archive

Against this backdrop, each volume in the series inquires into the multiple ways in which science and technology are translated, circulated, and become entwined with various other practices, a process which results in the making of heterogenous assemblages. Although the case studies examine different empirical events and situations, they all pay attention to the emergence and unfolding of ever-changing, yet patterned assemblages, while critically probing the analytical concepts they deploy.

The concept "assemblage," as used in much of the STS literature, is translated from the French word *agencement*, which has no exact English equivalent. Nevertheless, the word agency is easily recognisable in it. Agencement refers to an arrangement of heterogeneous elements adapted to each other, without assuming that a human agent has arranged them as a set of passive things according to a predetermined plan and in order to extend human agency. On the contrary, agencements are endowed with their own agency. For Deleuze and Guattari (1987) who coined the concept, there is nothing outside of an agencement, because its description and the construction of its meaning are part of it. This implies that a sociotechnical agencement includes the language that is necessary to describe it. In order to make this claim, Deleuze and Guattari have to contradict their own argument. They maintain that they can look critically at the agencement that they inevitably inhabit, while at the same time maintaining that this is impossible. But they are certainly in good company here, since it has been plausibly argued that human reason is capable of self-reflectively incorporating this dilemma (Wittgenstein, 1969: §86–115). To insist on this capacity of human reason is not to reaffirm the modernist claim that human beings are in control of the assemblages of which they are a part, nor to reject the opposite conclusion that human beings cannot be held responsible for the assemblages they have helped to create.

Assemblages, as I have emphasised above, are the result of a combination of heterogenous elements adapted to each other. Some of these elements are human and others non-human, some are material and others immaterial. One can first think of simple examples like a library, a hospital, or an airline. At the same time, and on a different level of classification, these combinations of heterogenous elements articulate differentiated social spheres—technoscientific, social, political, economic, legal, and cultural—with their distinct logics, codes, valuations, and normativities. These different logics, codes, valuations and normativities come together to form assemblages with their own agency, such as libraries, hospitals or airlines. Their agency does not lie solely in human design and practice, nor in the influence that one social sphere has on the others. Rather, the agency of assemblages arises from the way their elements interact.

So far, I have argued that assemblages and practices are co-constituted, that they are eternally evolving, and never set in permanently fixed patterns. The main cause for this never-ending change is prefigured in the fact that assemblages exist only as far as they are enacted in practices. Practices are patterned and thus imply repetition—otherwise one would not distinguish them from random activities and behaviours. But repetitions necessarily engender differences. What is repeated is at the same time being changed. Since practices are necessarily changing, they can bear a horizon of change for the better (Arendt 1998). One way to explore better possibilities for the future is to observe other people's practices and improve one's own. It is here that the notion of circulation becomes important for the approach that the authors of this series engage with. It is also here that it becomes clear why a search for origins fails to understand how assemblages work.

An assemblage cannot circulate, or travel, as it exists as a network of relationships between heterogeneous elements that co-constitute each other. What may travel are some of its disassembled elements. Think again of the three examples given above: library, hospital, airline—none of them can travel as such. But some of their constituent elements, such as books, medicines and airplanes, do. So can librarians, doctors and pilots, as well as cataloguing techniques, diagnostic equipment, and scheduling, repair and maintenance protocols. In order to depart, travel, and arrive, these elements must be disassembled from their assemblages and reassembled into new assemblages. This seems to imply that the travelling elements have a material core, which remains immutable during the journey, and a semantic form, which is transmuted during the journey. But the point here is that these two sides cannot exist without each other, nor without shaping each other. Thus, it would be wrong to assume that the materiality of the travelling element is in itself immutable, and the semantic form is in itself mutable. The material and the semantic are inextricably folded into one another, i.e. the semantic is not a representation or reflection of the material; together they are "translations" of each other. Therefore, when an element travels, there is nothing stable about it. Nothing is "given" when the element begins to travel from an assumed origin, and nothing tells us whether, after being disassembled and reassembled several times, we can still consider that the original element has endured. The answer to the question of the quality of a translation cannot be found in the original version of what has been translated. In fact, the belief that there is an original is an effect of translation, as we will see below. In short, the key operation that constitutes an assemblage and allows some of its elements to circulate is translation (Latour 1993; 1995).

Circulation, facilitated by translation, produces different kinds of "isomorphism" (things become similar), but over time and after several translations, the result is always a version of "transmorphism" (a new form emerges) (cf. DiMaggio and Powell 1983). Normative isomorphism is achieved through political regulation, as when all countries in the world agree to use the Gregorian calendar as a meta-code for regulating air traffic or stock markets, while continuing to use other calendars as their own cultural codes. Mimetic isomorphism is a form of learning achieved by adapting to the way others do things. This happens, for example, when many libraries (or hospitals, or airlines) around the world imitate a particular library (or hospital, or airline) that is perceived to be outstandingly successful. In such cases, a blueprint emerges and becomes a model through imitation. In other words, the model is not the original, but the product of imitation. Those who imitate a blueprint create the model. A third type is coercive isomorphism, where the blueprint is not imitated but people are forced to adopt it. These three types of isomorphism (normative, mimetic, coercive) are ideal types, since in most empirical cases two or three of them are at work simultaneously. In the end, they all produce new transmorphed forms.

The European colonisation of large parts of the world, including the African continent, did not begin as a project of coercive isomorphism. The intention was not to install any of the political, legal, and economic structures that were shaping

European countries at the time. On the contrary, colonial extraction worked better by avoiding isomorphism and often began with brute force, executed with superior weaponry. However, as European colonial powers pursued their economic interests, they were drawn into projects that implied coercive and normative forms of isomorphism in areas such as education, health, and basic infrastructure. These projects of normative isomorphism were accompanied by some forms of mimetic isomorphism on the part of the colonised, who tried to make the best of the situation, especially with a view to their liberation. We can say that European colonialism ended when a number of structures in the colonies—such as education, mobility, communication, and the form of the nation state—had become isomorphic enough to organise effective resistance and liberation movements that could drive out the colonisers.

For the purposes of this series, which focuses on the translation of technology in Africa, the relationship between the circulation of technology and transmorphism is the main problematisation. One of its important aspects is that the circulation of technology, which results in transmorphed forms, does not follow political borders or any other boundaries between, for example, cultures, religions or social spheres. The spatial extent of the resulting and constantly changing assemblages does not overlap with neatly delineated spaces such as national territories, regions, continents or social fields like for instance health care. Some infrastructures are designed to integrate a specific territory through enclosure, such as policing and border control, with their associated technologies of surveillance and movement control. Other infrastructures are designed to partially cut across national territories to integrate regions, such as roads, railways, airlines, telecommunication systems, riverbed management systems, and so on. And yet some other infrastructures are designed to span the globe, such as the Gregorian calendar, the base ten numerical system, the decimal metric system, and the Internet and its protocols. Finally, because social spheres have spatial dimensions, some infrastructures cut across them, others delineate their boundaries, and still others do both at the same time. For example, the UN Sustainable Development Goals Programme has created a global measurement network that delineates sixteen priority areas. The resulting measurement infrastructures produce indicators that allow comparisons between countries and regions, reveal patterns inside and across priority areas—such as between gender equality and poverty—and serve to also measure the impact of the programme. In other words, if circulation and isomorphism lead to new forms of technoscientific transmorphism, the latter creates its own spatiality.

It is important to note that this argument has a temporal dimension. As new technologies emerge—through transmorphisms—old ones fall out of use and become part of technological archives. A technological archive may be defined as a particular form of cultural memory that contains a range of obsolete technoscientific devices and procedures, many more than any individual or collective actually remembers, but which can be unearthed—for example, ancient Babylonian wheat seeds kept in seed banks that are now proving useful for current climatic conditions. To assume that there are two clearly delineated archives (North/South), or five (continents), or two hundred and six (countries), would be to misunderstand the argument of archives as epistemic spaces that cut across countries and history.

To reiterate, the contributors to this series explore ways in which the circulation and translation of technoscientific things—e.g. units and formulas of measurement and statistics, technologies, networked infrastructures, methods of approximation, procedures and tools of data collection and analysis—produce what are here called isomorphisms and transmorphisms (without them necessarily using this vocabulary). The hope is to contribute to a body of knowledge interested in future-making and to provide tentative answers to the following question: How can technoscientific knowledge production become part of the *commons* to serve a just and sustainable future for all people and the planet?

3 Probing Themes: Metrics, Technicisation, Infrastructures, Technoscapes, Devices

The volumes in this series examine the unfolding of sociotechnical assemblages, in the present as well as in the past, and in relation to the following themes: metrics and digitalisation; technicised lifeworlds; infrastructures; technologies and space-making; devices and their users. Their shared aim is to support, provoke, and stimulate emerging debates on the operation of technosciences on the African continent and, more generally, through global interconnectedness.

While these debates are, of course, characterised by unresolved controversies, the series brings together detailed empirical work and theoretical reflection to raise new questions. We do not pretend that we can resolve the ongoing controversies and answer with certainty the open questions raised in this introduction. We certainly do not believe that we can create a metaphysical tabula rasa to craft a new ontology or unearth an old one. Rather, we adhere to Otto Neurath's interpretation: "There is no tabula rasa. We are like sailors who have to rebuild their ship on the open sea, without ever being able to dismantle it in dry-dock and reconstruct it from the best components. Only metaphysics can disappear without trace" (Neurath 1983, 92).

Bibliography

Arendt, Hannah. 1998 (1958). *The Human Condition*. Chicago: University of Chicago Press.

Chalmers, Alan F. 1976. *What is this Thing called Science? An Assessment of the Nature and Status of Science and its Methods*. Queensland: University of Queensland Press.

Deleuze, Gilles and Félix Guattari. 1987. *A Thousand Plateaus: Capitalism and Schizophrenia*. Minneapolis: University of Minnesota Press.

Diagne, Souleymane Bachir. 2013. "On the Postcolonial and the Universal." *Rue Descartes* 78, no. 2: 7–18.

DiMaggio, Paul J., and Walter W. Powell. 1983. "The Iron Cage Revisited: Institutional Isomorphism and Collective Rationality in Organizational Fields." *American Sociological Review* 48 (2): 147–160.

Harding, Sandra G. 2011. *The Postcolonial Science and Technology Studies Reader*. Durham N.C.: Duke University Press.

Hountondji, Paulin. 1990. "Scientific Dependence in Africa Today." *Research in African Literatures* 21, no. 3: 5–15.

Latour, Bruno. 1993. *We Have Never Been Modern*. Cambridge, Mass.: Harvard University Press.

Latour, Bruno. 1995. "The 'Pédofil' of Boa Vista. A Photo-Philosophical Montage." *Common Knowledge* 4, no. 1: 144–187.

Mbembe, Achille. 2021. *Out of the Dark Night: Essays on Decolonization*. New York: Columbia University Press.

Neurath, Otto. 1983. "Philosophical Papers, 1913–1946." In *Vienna Circle Collection*, edited by R. S. Cohen, Marie Neurath, and Carolyn R. Fawcett, no. 16. Dordrecht: D. Riedel Publishing Company.

Rodney, Walter. 1972. *How Europe Underdeveloped Africa*. London: Bogle-L'Ouverture Publications.

Stengers, Isabelle. 2015. *In Catastrophic Times: Resisting the Coming Barbarism*. Lüneburg: Open Humanities Press and Meson Press.

Suchman, Lucy Alice. 2007. "Reconfigurations." In *Human-machine Reconfigurations: Plans and Situated Actions*, 259–86. Cambridge: Cambridge University Press.

UNESCO. 1974. *Science and Technology in African Development*. Paris: The Unesco Press.

Verran, Helen. 2002. "A Postcolonial Moment in Science Studies: Alternative Firing Regimes of Environmental Scientists and Aboriginal Landowners." *Social Studies of Science* 32 (5–6): 729–762.

Anderson, Warwick. 2002. "Introduction: Postcolonial Technoscience." *Social Studies of Science* 32 (5–6): 643–658.

Wittgenstein, Ludwig. 1969. *On Certainty*. Oxford: Blackwell.

CHAPTER 1

On Technicisation: How to Create a Zone of Decolonial Translation?

Richard Rottenburg and Eva Riedke

1 The Argument

The problematisation that this volume undertakes is best outlined from the outset. The "question of technology," which has haunted modernity since its inception, is in the current moment being controversially reframed in the light of the real possibility of an ecological, political, and social collapse of the planet. In some regions of the world and in some of their social worlds, an atmosphere of finitude has become the new spectre of the contemporary. These social worlds no longer ask what technology is. The question has become more practical: Who, what, when, where, why, and how should the role of technology in shaping the planetary future be changed?

At the latest since the beginnings of European modernity in the 18th Century and its forced expansion throughout the world to generate multiple versions of itself, technology has increasingly manifested itself as a realm of mechanisms that operate according to means-end schemata. Modern technology is designed to increase performance, efficiency, and acceleration, and to thus enhance human capabilities. We can say: technology has become a design for the infinite that seems to compensate for the finite human capacity in terms of time, space, and knowledge. This design has always provoked multiple radical questions. One of the early critical questions has been: Where, in the face of irreversible human finitude, does the drive for infinite increases in performance emanate from?[1]

At the end of the 20th Century, when humanity had to realise that it had been living for some time in the Anthropocene or Capitalocene, the fundamental question of the aporetic drive of modern technology took on a more urgent

---

[1] The most influential thinkers who have shaped this question have come from those regions of the world where technicisation and industrialisation, and the problems associated with them, were most advanced at an early stage. Many of them wrote in German and include Karl Marx, Edmund Husserl, Martin Heidegger, Herbert Marcuse, Max Horkheimer, Walter Benjamin, Hans Blumenberg, Reinhardt Koselleck.

significance. In the West, the general acceptance of this realisation came rather late, even though the role of technology in co-producing the catastrophes of modernity had already become apparent to some observers in the first half of the 20th Century (Benjamin [1939] 1969). For those who became victims of its technicised reach for the infinite—first in Europe (Marx [1858] 1983) and very soon throughout the world—the primary question was not where this drive came from, but how to resist or escape it? (Fanon [1961] 2004). As doubts gradually accumulated over the centuries and intensified in the second half of the 20th Century, the question of technology was reformulated: Do we need to change the design of modern technology to leave behind the endless increases in efficacy, efficiency, and acceleration? By the end of the century, the numerous controversies generated by this question had finally been settled at least on the most general level: Yes, we need to change it! (Stengers [2009] 2015; Sloterdijk 2023). At the beginning of the 21st Century, the remaining and difficult part of the question concerns the modalities of how to do it.

Accordingly, the work presented in this volume focuses on the "who," "what," "where," "when," and "how" part of the fundamental problem. Within this vast subject, the volume is guided by the sense of finitude, urgent repair and decolonisation that have become core challenges of the present. Through empirical investigations, the volume seeks to provide fine-grained praxiographic material that prompts us to ask new and specific questions related to its concrete object of investigation: technology.

The six empirical chapters in this volume examine different modern networked technologies in several contemporary postcolonial African countries. They empirically explore the provision of drinking water and electricity in Kenya, telecommunications in Niger, health care and policing in Uganda, and civil registration in Ghana. The authors examine the design, operation, use, materialities, politics, economic dimensions, and meanings of these technologies, asking how they affect ways of ordering life and, conversely, how they are affected by existing and always multiple and changing social orders. In doing so, the authors approach the analysis of governance through technologies and ask how they are intertwined with both the lifeworlds of accountable organisations and the everyday lives of the citizens whom they are meant to serve and govern. In light of these main lines of inquiry, each chapter focuses primarily on the ongoing practices of translating circulating technologies into various lifeworlds, in selected African countries, under particular circumstances. This kind of praxiographic work opens up new spaces of analysis and action which, in turn, so our hope, render tangible possibilities for a decolonial appropriation of technology.

In order to make the arguments presented in this volume more comprehensible to the reader, a note on our position with regard to modern technosciences seems helpful (see also Beisel, Calkins, and Rottenburg 2018). In our view, there is ample evidence that climate change is caused by industrialisation and facilitated by technoscientific developments driven by capital interests. However, we would immediately recall that climate change was predicted half a century ago by technoscientific efforts. While this was only considered by the minority of scientists in the early 1970s, by the mid 1980s the majority of the scientific community deemed the effects unquestionable. They joined forces to try to persuade governments to change their policies on fossil fuels. Without technoscience, we would still not know the causes of extreme weather events increasingly being experienced around the world. Similar arguments can be made about zoonotic diseases and pandemics, highly uneven population growth, poverty, hunger, and agricultural production. Technoscience is complicit in the forces that are destroying the planet's biosphere. And it is indispensable for finding ways out. All the related problems are interconnected on a planetary scale. Their solutions must therefore also be planetary in scale.

Having set out the guiding question for the volume, the next section will introduce our theme by providing a brief outline of the work and the empirical research that underpins the volume. We then systematically develop our argument in a series of logically connected steps, presented in three hermeneutic circles (sections 1, 2, and 3), in which the key concepts and their interrelations are introduced in order to construct the theoretical scheme that guided the empirical work and the interpretation of the praxiographic material collected by the authors.

In the first hermeneutic circle (1), we will focus on the concept of "translation". We will begin by elaborating the translation of "travelling technology," then show the link between translating technology (as opposed to transferring technology) and "routinisation," in order to close this circle by zooming in on the role that "technological archives" have for translation and "innovation".

In the second circle (2), we will go more into depth regarding these, by now, familiar concepts and their interdependence in order to develop our theoretical scheme. Here we move first from routinisation to an explanation of the terms "lifeworld" and "technology," in order then to approach our most important key term, "technicisation," and its constitutive connection with lifeworld.

In a third and final hermeneutic circle (3), at the centre of our argument, we will discern two aspects at the bottom of the lifeworld-technicisation entanglement: technicisation in meaning-making and technicisation in decolonisation. On the basis of this interpretive scheme, we will then be able to present the case studies a second time, now in greater depth. The final section will draw together a number of our conclusions.

2   Introduction

Between 2011 and 2019, the contributors to this volume, together with many other colleagues, worked on a research programme studying transformations of technologies and their significations in Africa (DFG 2019). The researchers came from a variety of disciplines in the humanities and social sciences, and focused on the creative and adaptive capacity of institutions and other actors to respond to the challenges of the postcolonial era. Within this framing, the guiding question was: How do creative adaptations of technologies produce new sociotechnical figurations for coping with material, economic, political, social, and environmental challenges? The search for answers was inspired by two core assumptions that seem uncontroversial in some academic circles, but continue to be ignored in others, especially by people in applied fields and in particular those concerned with technology transfer.

The first assumption of the programme was that no circulation of ideas, artefacts, or technologies could possibly have a single origin. It is only through crude simplifications that technological and scientific achievements can be attributed to a time and place of origin, sometimes even to an inventor. But none of these achievements—such as the insight of Galileo Galilei, Nicolaus Copernicus or Muhammad ibn Musa al-Khwarizmi (van der Waerden 1985)— would ever have been possible without an infinite chain of previous achievements, usually made long ago and in distant places. To make this point, Robert K. Merton famously rephrased the Latin metaphor for influential thinkers from "dwarfs standing on the shoulders of giants" to "dwarfs standing on pyramids of other dwarfs" (Merton 1993). Michel Serres ([1989] 1995, prologue) translated this idea into a spatial image, arguing that the great inventions of mankind are best compared to great rivers, which only come into being because there are an almost infinite number of small springs upstream. These join together to form countless small streams, which in turn form many rivers, which gradually join together to form the great river that flows to the sea.

From the 18th Century onwards, imperialist Europe perceived and presented itself as the sole source of all superior knowledge, even though it was obvious that knowledge emerges from many sources and that these are spread across the different continents. In attempting to correct this deception, it is easy to mistakenly confirm European claims to be the origin of all advanced knowledge. This happens either when one rejects valued knowledge as "Western" or "Eurocentric" rather than questioning the imperial myth of its European origin. Or it happens when one falls into the inverse but logically equivalent problematic trap of claiming exclusive intellectual property rights for the same knowledge by pointing to a different, yet again single origin. Both forms of argumentation distract from the more reasonable conclusion that all great creations belong to humanity as a whole.[2]

The second core assumption of the programme states that new sociotechnical figurations necessarily result from the flow or circulation of ideas and technologies (Appadurai 1996). This implies that circulation is made possible by translation.[3] In this sense, creative adaptations were conceptualised as practices of translation that take place in the interstices between different fields— social fields, discourses, orders of justification, between different time periods and between communities, countries, regions, and continents—and connect them in novel ways. Accordingly, one would need to pay particular attention to the role of technology in translation practices. The precondition for the possibility of translation is the existence of a translation regime, i.e. an archived

2   See several interesting case studies in Serres ([1989] 1995); on the history of mathematics see also Diagne (1989). Achille Mbembe looks at the dark side of this phenomenon. Europe has not only had the greatest power to absorb more knowledge from around the world than any other part of the world, it also had the greatest power to radiate and spread the ideas it absorbed from around the globe as its own ideas. Mbembe borrows this observation from Paul Valéry, but develops it with the help of Giorgio Agamben's reflections on biopolitics and the figure of the camp. He suggests that Europe's radiant power stems from the singular European idea that it has a biologically based mission to bring the entire world under its biopolitical control. The consequences of this are colonialism, the figure of the camp, and racism (Mbembe 2016, 98ff). We prefer to leave the teleological dimensions of such arguments to the philosophers of ideas and stick to our focus on the contingency of constantly unfolding assemblages. These, we argue, are not driven by ideas alone, and the implied ideas do not remain the same as they travel (see below).

3   One of the authors of this essay has been working and publishing on translation (as the term is used here) for many years: Rottenburg (1994, 1996, [2002] 2009, 2003, 2014; Kaufmann and Rottenburg (2012). See a very close application of the term in von Schnitzler (2016). As the term is understood here, it implies an approach to the definition of "universal" beyond its misuse, see Senghor (1956); Diagne (2013).

stock of basic assumptions and patterns about how to recognise the unfamiliar as potentially useful. In our case, the question is how to recognise a new technology as potentially useful. Technology involves a means-end scheme and is entangled with material, economic, social, political, legal, cultural and aesthetic aspects, and aims to produce a particular order of things. It is therefore inevitably normatively charged (Bijker, Hughes, and Pinch [1987] 2001).

More specifically, the research programme has shown that technologies and their significations—the meanings and sets of rules that are always inscribed in them—can only be understood through their entanglements since there is no such thing as pure technics or pure significations. Finally, the research showed how such entanglements are partly dis-entangled and re-entangled in the course of their travels and circulation. In short, for a technology to circulate, it has to be translated, and in the process of being translated, it changes as it enters new assemblages. In other words, it acquires new meanings and new functions.[4]

Most importantly, it has been shown in great empirical detail that, at the most critical point of translation, it is crucial to establish a link between the circulating technology and the technological and institutional environment into which it is being translated. Otherwise, sooner or later, the technology will be dropped and disappear again. In the context of the problematisation chosen for our research work, the resulting question was: What distinguishes a good translation from a bad one? We were able to show that an answer can only be found by critically considering normative aspects in the analysis. It should be noted, though, that normativity is not brought into play by the researchers, but is inevitably already inscribed into the respective technology and is actively invoked in the local processes of translation (Akrich 1992). In redefining the use of technologies, actors are guided by multiple and sometimes conflicting value systems and justifications, which entail elementary assumptions about what their self-interest is and how it relates to the common good. These value systems necessarily change in the process of translating a new technology, and for this reason the study of translation offers privileged insights into the workings of power asymmetries and hegemonic orders of justification.

In short, the means-end schemata built into technologies changes through circulation and translation in such a way that new meanings and purposes emerge and old ones are erased or fade into the background. The new purposes may or may not be desirable for all those involved and affected by the process. Ensuring the participation of as many stakeholders as possible—often

---

4   The notion of assemblage is used in this essay as it was introduced by Gilles Deleuze and
    Félix Guattari ([1980] 1987) and later specified to be more useful for empirical analysis by
    John Law (2008). Further down we elaborate on the relevance of assemblages in more detail.

discussed under the concept of lay or citizen science—certainly improves not only the legitimacy but also the quality of the translation process. While lay participation has its obvious limitations when technoscientific expertise is not easily translated into common sense, it is indispensable when trying to prevent technoscientific expertise from serving political power. Translating a circulating technology into an existing web of technologies, institutions, and beliefs usually challenges inherited understandings of individual and group interests, and invites redefinitions of both. The resulting controversies over what was previously taken for granted can only be resolved through negotiation and compromise, since the criteria for distinguishing arguments that are more convincing from those that are less convincing are also in flux. In many cases, however, some of the participants in these negotiations are global actors with significantly more power and resources than most others. Our research on the translation of circulating technologies has therefore paid particular attention to the many ways in which power asymmetries influence compromise.[5]

The chapters in this volume are intended to extend the research programme outlined so far. They were selected for this edited volume because of their interest in detailed analyses of translation practices that involved circulating technologies. They all pay careful attention to the ways in which circulating technologies are translated into the networks of technologies and institutions that already exist in the places into which they are translated. In pursuing this focus, the authors in their analyses point to the workings and intertwining of two emancipatory agendas: the decolonisation of technology, and the decolonisation of postcolonies. Both agendas are highly present in current public and academic debates, bringing with them an immediacy and intensity in terms of the expectations of political relevance that characterises research in the contemporary moment.

One emancipatory agenda is to regain human control and responsibility over technological development. This is based on the widespread assumption that technoscience no longer serves human purposes, but rather undermines them. For the sake of clarity, it can be said that the aim of this agenda is to decolonise technology. The second emancipatory agenda is about reclaiming self-determination and the betterment of postcolonial countries. Its aim is to advance the decolonisation of the postcolonies as a political project. The two agendas seem to continue to coexist as separate, particularly at the level of discourse and political rhetoric. On closer inspection, however, they are, as we argue, deeply entangled. This becomes apparent when one realises that the

---

5　For the general argument see Law (1991); for a particular case see Appel (2019).

second of the two agendas, namely that of political decolonisation, comes in
two versions with opposing aims. One emphasizes independence through the
transfer of cutting-edge technoscience for a sustainable future for Africa (see
African Union 2023). The other version sees the transfer of technoscientific
models primarily as a gateway to perpetuate old colonial dependencies and
create new ones—imported technology is likely to be another Trojan Horse
(Adas 1989). In this volume we argue that the apparent contradictions between
these two versions of the second agenda are the result of flawed assumptions
and can be resolved by paying closer attention to the translation of circulating
technologies.

The chapters in this volume have been written with the aim of exploring the
entanglements between the two agendas, or, in other words, the ambivalent
entanglements between technicisation and colonisation. For example, Jonas
van der Straeten and Jochen Monstadt analyse the strategies of the utility company Kenya Power to adapt the provision of electricity to the social structures
of different parts of Nairobi, emancipating itself from the global model of how
to stabilise electricity distribution in large cities. Jannik Schritt examines the
creative translation of text message chains into political mobilisation in Niger
during an uprising in 2011. Sarah Biecker, Jude Kagoro and Klaus Schlichte
examine the different levels of normalisation of three different technologies of
government (the police file, the breathalyser in traffic policing, and accounting
procedures in the context of national budget planning). Alena Thiel examines
the introduction of biometric identification technologies in Ghana and shows
how this process implies both a technological consolidation of inherited Ghanaian classifications of human kinds and an increased level of critical questioning of their hegemonic character. The other two chapters follow the same
interest in the entanglement of the two agendas, but focus more on the asymmetries within the translation process (see below).

In order to examine the entanglement of the two agendas, we argue that
the following key questions need to be asked: How do technological archives
and their regimes of translation shape the creative adaptation of a new technology? How do certain technologies come to be taken for granted? Which
of the multiple meanings of a technology become self-evident? Is this self-
evidence linked to the creation of new inequalities? What kinds of worlds are
produced and what kinds of worlds are repressed by different forms of technological change? How do we assign human responsibility for the effects of a
process that seems largely beyond human control? And how does all this relate
to decoloniality? Given these questions, the modus operandi of the empirical

investigations of the chapters is based on the primacy of praxis and thus on praxiography (Akrich 1992; Star 1999; Bueger and Gadinger 2018).

In 2017, Clapperton Chakanetsa Mavhunga, a historian of science and technology at MIT, published the edited volume What Do Science, Technology, and Innovation Mean from Africa?. He begins the introduction by taking a number of positions, some of which we share and some of which we do not (Mavhunga 2017, 1–9). While his interest is mainly in unearthing historical forms of African science and technology, ours is more general and mainly asks contemporary, forward-oriented questions. Nevertheless, he ends the introduction with a question similar to the one we developed for our research programme back in 2011. Like us, he ultimately wants to know more about the workings of creative resilience and how it can be supported in the face of pressures to adapt to economic and political orders and imported technologies. To make his point, Mavhunga (2017, 9–21) uses some snapshots from precolonial, colonial, and postcolonial history. Where we speak of "creative adaptation" and, at a deeper level, "translation," Mavhunga (2017, 18) speaks of "imitation" complemented by "the creation of synergies between inbound and locally invented modes of innovation and entrepreneurship". We thus share the conviction that much more can and must be discovered about how exactly these "synergies," to use his vocabulary, work in order to prevent the ongoing destruction caused by the blind importation of "inbound" technologies. And, more importantly, in his 2018 monograph, he provides invaluable historical material and insights needed to reconstruct African technological archives that are key to what we have in mind when we set out to explore translating travelling technologies and the innovations they lead to (Mavhunga 2018).

Unlike Mavhunga, and perhaps less humbly, we locate this common interest in a different and larger cause: We ask how the entanglement of lifeworld and technology might be conceived and designed in order to create spaces for the decolonisation of technology? More specifically, we ask what it might mean to conceive of the translation of circulating technologies as a decolonial practice aimed at liberating technology from its design for infinite increases in performance and acceleration within the framework of capitalism and modernity? And what would it take instead to adapt technology to the given and finite human condition? We believe that our approach supports Mavhunga's main concern to unearth successful local forms of practicing science and technology.

The challenge that arises here is that our argument also supports that of Julie Livingston, who makes a U-turn in the opposite direction. In her magnificent monograph Self-devouring Growth (2019), she exposes with unrelenting directness the devastating effects of a whole range of imported technologies. She meticulously demonstrates how these were not creatively adapted to local needs and therefore resulted in autophagic growth. In the language of our essay, Livingston shows how the pursuit of infinity can destroy a world at peace with its finitude. Our aim is to develop an interpretive scheme that neither reduces the contradiction between Mavhunga's and Livingston's studies to something that results from the authors' differing inclinations, nor ascribes it to an error on either author's part. Rather, we want to show that these are the extreme ends of a spectrum of possibilities that can be traced back to more or less successful translations of circulating technology. The empirical chapters in our volume do not pose these ambitious and fundamental questions directly, but rather interpret their empirical material against the background of the concerns outlined here.

In the next section we enter the first hermeneutic circle to unpack the deeper layers of the concept of translation. We begin by elaborating on how we come to understand the translation of travelling technology, then show the link between translating technology and routinisation, and finally close the first circle by zooming in on the role of technological archives for translation and innovation.

3   Translation

3.1   *Travelling Technologies and Translation*

In short, the starting point of our analysis is the practice that enables the confluence of knowledge flows. We understand this practice as translation, and its site is the encounter. In idealised terms, the encounter takes place between a circulating idea, necessarily tied to a material carrier, a specific context, and a conducive situation. The situation is often characterised by a disruptive event—for example, an epidemic such as AIDS, Ebola, Zika—followed by reorientation efforts aimed at increasing receptivity to circulating technologies that could be adapted to change the situation for the better. This practice requires a kind of creative adaptation of both the receiving context and the circulating technology.

In most uses of the word adaptation, it has one of two different meanings, both of which imply external pressure. It can mean adapting to overwhelming circumstances by discarding existing structures and replacing them with supposedly superior ones. This may begin as resigned submission, but after a while it may continue out of deep conviction and even messianic fervour. Or it begins as a tactic aimed at preserving existing structures and epistemological and normative orders in secret, behind a facade of adaptation to a new situation. After a while, however, the facade can become a matter of course, the original motif is gradually forgotten, and a real adaptation takes place.

Creativity, on the other hand, in most uses of the word, refers to a practice that aims to change things on the basis of better insights rather than external pressures. These better insights are understood to be the result of a special talent that is difficult to name or to influence. While adaptation to circumstances is seen as a form of passive submission, creativity is at the opposite end of the same semantic spectrum. Here we find inventions made out of free will, out of the joy of play, or out of intelligence that is not consumed by the struggle for survival.

This discursively widely consolidated juxtaposition of adaptation and creativity has far-reaching consequences for the distinction between what is today understood in a problematic binary as the Global South and the Global North (mostly still meaning "the West"). This binary is embedded in flawed notions of modernity, progress and development. According to it, some regions of the world and their peoples seem to be trapped in the role of perpetual backwardness, trying to catch up with the development of other regions by leapfrogging. This in turn means that whatever they do, they can only imitate something that has already been invented in more advanced countries or social fields. The worst denigration that can result from this ideology is to interpret an existing modern practice as an imitation and cover-up of an older, underlying, non-modern practice.

However, if we think of adaptation and creativity not as opposites, but as the two extremes of a continuum of creative adaptation, a new perspective on the potential of translating circulating ideas and technologies opens up. On this assumption, it seems plausible that the aim of translation cannot be the faithful reproduction of a circulating technology. It must be its creative modification so that it can function in a new and different context. Translating a travelling technology is therefore less a matter of preserving its form than of achieving a comparable effect. The practice of creative adaptation is guided by regimes of translation that are part of technological archives, which in turn distinguish one technological zone from another and thus never fully correspond to countries, continents, religions, languages, or fields of expertise and practice (explained in the next section). The translation of circulating techno-scientific models is therefore an achievement that takes place between several technological zones. This is where most of the surprising discoveries are made. In light of this observation, the false juxtaposition of inventors and imitators disappears, and with it the categorisation of the former as belonging to the

Global North and the latter to the Global South. What emerges is a space of serendipity, which is the best condition for creativity everywhere.

A common view related thereto is that excellent resources and freedom from immediate material needs are prerequisites for creativity. In this view, creativity arises out of satiation, sometimes boredom, and the desire to rearrange things out of playfulness, curiosity, and experimentation. An opposing view insists that deprivation and lack of resources are the better conditions for creativity. These two ideas can be geographically very close. Silicon Valley is seen as the centre of innovation because of the immense resources available there. On the other hand, the deprivation theory of innovation is also celebrated within the very same Silicon Valley tech community. Important inventions are often said to have been made by young, formally unqualified, and initially unknown individuals with astonishing skills and little capital in backyard garages, i.e. on the fringes of well-equipped companies and research institutions.

Like the juxtaposition of adaptation and creativity, the juxtaposition of abundance and scarcity is similarly imprecise and misleading. The relationships are more nuanced, but also more prosaic. To create something new that is meaningful to other people requires all the resources necessary to do so. Without access to digital devices and an internet connection, one cannot participate in software development. But it is safe to assume that experimentation motivated by saturation and boredom is less determined and indomitable than experimentation motivated by an existential problem. Accordingly, there is no reason to expect that innovation will come mainly from the South or mainly from the North. In the case of digital technology, this observation is particularly true in that, for the first time in the history of technology, the infrastructural conditions for an innovation may well be on the other side of the world—in Silicon Valley or Shenzhen, for example—because a high-speed fibre-optic connection may be sufficient to create similar conditions wherever you are. From this perspective, the crux of this still young world of global digital connectivity is the extent to which it will either perpetuate or exacerbate old patterns—with peripheral locations having to adapt to the software of metropolitan locations—or, conversely, create the conditions for future generic software to be developed differently. More specifically, it is about creating generic software that, while inevitably forcing global standardisation, can still bridge differences between spaces of use without the colonising effects of past technological standardisations (Pollock, Williams and D'Adderio 2016).

The terminological apparatus presented above is not intended to solve the underlying problem caused by the persistent asymmetry between those who collaborate in international projects to make a technology work meaningfully in African contexts. The terminological apparatus is intended to develop an analytical lens to look for solutions in the right place. The hypothesis presented here is that in order to find solutions, one must first identify and strengthen zones of decolonial translation. We argue that there are niches where decolonial translation has always taken place, and the chapters in this volume contain several stories of this. But these same stories also carry another leitmotif, which is concerned with the persistent postcolonial asymmetries that characterise, to varying degrees, the translations examined in the chapters. While all six empirical chapters deal with this leitmotif, two of them take it as their central theme. The chapter by Christiane Tristl and Marc Boeckler shows how a water dispenser designed in Europe and introduced in rural Kenya is deliberately designed to prevent local creative translations as far as possible. The automated public water dispenser can be seen as an iconic counter-example to the famous Zimbabwean bush pump (de Laet and Mol, 2000). Sung-Joon Park's chapter shows how the huge and expensive international programme to provide access to HIV treatment in Uganda is structured in such a way that potential translators are largely prevented from creatively adapting the system to the local situation, simply by not being given sufficient project time to do this essential work.

It seems important to emphasize here that finding better ways of dealing with or eliminating the asymmetry in this step of translation (the interaction between local and international experts) does not automatically mean that subsequent steps of translation (between the social fields of a country) will necessarily be successful. Other asymmetries, inequities and problems will arise when the technology is translated between the social fields of a country, e.g. between social classes and between economic, legal, and cultural considerations (some of which are explored in the empirical chapters of the volume, see in particular the chapters by van der Straeten and Monstadt and by Schritt).

Agreeing with Mavhunga (2017, 5), we find it striking that over the past three or four decades, relatively few scholarly contributions from and about Africa have focused on technological translations that work. The majority have instead focused on the disruptions, decay, brokenness, accidents, and risks associated with the workings of technology on the continent. A number of contributions have sought to understand what the frequent disruptions and extended states of decay really mean for the understanding of technology in general.[6] This problematisation is of particular concern and interest when networked technologies fail to function as the reliable background condition that

---

6 See for instance Simone (2004); Trovalla and Trovalla (2014); Larkin (2013); Furlong (2014); Anand (2017); Anand, Gupta and Appel (2018).

modernity promised for economic development, public institutions, and private life (Law 1994; Star 1999).

By foregrounding the broader historical context of this phenomenon, most influential studies have shown how technologies are entangled with colonial histories, and how they later become part of postcolonial development trends. These established studies use the term colonisation to refer to the subjugation of large parts of the world by European colonialism.[7] Our argument does not question the findings of these studies. Rather, it builds on them and seeks to add a future-oriented perspective to strengthen them. To achieve this, we propose an analysis of how the understanding of colonisation in much of the established literature can be made more forward-looking by combining it with an understanding of colonisation that is usually attributed to the work of Jürgen Habermas ([1981] 1989). He argued that modern technocracy colonises the lifeworld and conceived of this as a universal problem. Without necessarily using the term, the understanding that forms of life and social organisation are transformed by technocracy has a longer genealogy, dating back to its theoretical centrality in the Frankfurt School of Critical Theory, long before Habermas (1968) began writing on the issue. The core assumption was that both forms of colonisation aim at subjugation, exploitation, and alienation, are linked to the mechanisms of capital reproduction, and generate various forms of mostly structural violence.

We are interested in the interrelationship between these two forms of colonisation, the kinds of inequities they produce, and the possibilities they might open up for decolonial translations. Before proceeding with this particular layer of our argument in the second hermeneutic circle, it is necessary to situate our approach briefly in the context of the available literature, which is closer to the objects of our research and has sought to investigate forms of routinisation.

3.2   *Translation and Routinisation*
The authors in this volume address questions that have been raised in similar forms, mainly in three now classic STS approaches: the Social Construction of Technology (SCOT), Large Technical Systems (LTS), and Actor-Network Theory (ANT). Building on this body of literature, the authors argue that technology and society are not ontologically distinct entities. In this view, the

---

7   See for instance: Rodney (1972); von Albertini and Wirz ([1976] 1982); Amin ([1988] 2010); Hountondji ([1983] 1990); Osterhammel (1995); Bayart (2000); Cooper (2002); van Laak (2004); Gandy (2006, 2014); Björkman (2015); Breckenridge (2016).

technical is not the material basis of the social but the two are relational and co-constitutive.[8] Accordingly, society is an inherently sociotechnical assemblage. To understand technology, one must look at the practices of doing sociotechnical assemblages. This position asserts that technology is not constituted as a set of devices designed for a purpose freely determined by humans. The social does not determine the technical, nor does the technical determine the social. It rather asserts that political and economic mechanisms—such as markets, neoliberal logics of regulating access and supply, or the power of states or corporations—shape, but do not determine, the kind of technology that emerges. Like the material, economic, political, legal, social, and aesthetic, the technical has its own logic. Agency is distributed between these different logics, and none of them alone are enough to make the world go round. They spawn assemblages with distributed agency that are continuously unfolding and remain forever open-ended. This also means that Latour's phrase of the early 1990s, "technology is society made durable" (Latour 1991, 103–21), makes a valid point, but misses two other points: that technology has its own logic and that a more technicised society is no more indestructible nor resilient than a less technicised one.

Within this broad argument about the impossibility of invoking the technical without the social and vice versa, questions of routinisation, understood as the result of the entanglement of the social with the technical, are key for our purposes. Some aspects of routinisation have been explored in the literature mentioned above, using five main terms that have become largely undisputed language of the STS vocabulary. "Black boxing" was introduced as a key term in Actor-Network Theory (ANT) and later reformulated by material semiotics. According to Latour, black boxing is a term that captures how successful technoscientific devices are designed to draw attention to their output and make their own internal complexities and vulnerabilities invisible (Latour 1999, 304; Law 2008, 8). The second term, "invisibilisation," refers more to the functioning and effects of infrastructures and large technological systems (LTS). According to Susan Leigh Star (1991, 381), users usually understand an infrastructure as "transparent to use, in the sense that it does not have to be reinvented or assembled for each task". But how an infrastructure actually works remains

---

8 This is a different relationality than the one implied in the definition of infrastructure—like for instance the power grid that serves as an infrastructure for computers, which then serve as infrastructure for the internet and, in turn, is the infrastructure for the operation of stock-markets, and so on in endless loops. For an ethnographic illustration of this latter notion of infrastructural relationality and an engagement with the practices of "suturing together" different forms of infrastructure, see Riedke and Adelmann (2022) and Riedke (2023).

largely invisible and mysterious to most users. The invisible qualities of a functioning infrastructure, in turn, become visible when it succeeds unexpectedly
(see Schritt in this volume), or when it fails or is tinkered with (see van der
Straeten and Monstadt in this volume).

The third term, "closure," refers to a particular detail in the interaction
between designers and users of technology that leads to closure by black boxing. The development of the automobile has become a classic in SCOT literature. The Ford Model T was first sold in 1908 as an "open" technology, inviting
users to modify it for different purposes (such as driving on roads, ploughing
fields, using the engine to power other devices). With each new model, however, the manufacturer closed the technology more and more, to the point
where today the user of a modern Ford can essentially change nothing (Kline
and Pinch 1996; see also the chapter by Tristl and Boeckler in this volume). The
fourth term, "inscription/description" (Akrich 1992), emerged from debates
within ANT and the related material-semiotic approach. Inscription emphasizes the norms, values, and ideas that designers believe the expected users
will or should have. Designers therefore inscribe them into the technology to
encourage certain uses and discourage others. De-scribing, in turn, highlights
the ways in which actual users deploy, translate, and appropriate a travelling
technology in the context of their lifeworld and technological archive, in part
by re-inscribing it with other meanings and forms of use. The process of de-
scribing and re-inscribing thus continues until the technology generates and
stabilises a new routine that will endure for some time. To these terms with
closely related meanings, each capturing a different aspect, John Law (2004)
added a fifth term when he introduced the notion of the "hinterland" of a
technical object or process. The hinterland refers, we might say, to the archive
of implicit presuppositions for doing certain things technically. For a presupposition to be in the hinterland means that it is both taken for granted and
hidden in the cultural subconscious. Past alternatives for the present, uncho-
sen and largely forgotten, and present alternatives for the future are almost
unthinkable when they challenge those understandings of their time that have
become so deeply rooted in the hinterland that they seem to express the essential nature of things.

For the purpose of our argument, the most important aspect of what all
five related terms seek to capture relates to the practice of translating circulating technologies. As explained above, for a technology to travel, it must
first be translated into a new context and then entangled with existing networks of technologies and institutions. It is here, at the point of entanglement, that de-scription and re-inscription take place. It is here that a
travelling technology can become a matter of course in a lifeworld with

its routines.[9] This step of translation is possible, we argue, because there is a register of translation from which the indexes for creating equivalences can be drawn. We further argue that this register is extracted from the technological archive of the people and institutions involved in translation. In another vocabulary, we would say that the register of indexes is part of vernacular knowledge. We are primarily interested in those moments of translation which are not predetermined by the conditions under which they take place, but which involve a moment of free action—including disputes, negotiations and resulting responsibilities—in the sense that Hannah Arendt ([1958] 1998) has given this term. This claim will have to be elaborated in the next step, which is the last section of our first hermeneutic circle.

3.3   *Innovation, Translation, and the Technological Archive*

The emergence and diffusion of new technologies cannot be explained in terms of their superiority over other ways of doing things. The logic of this argument is as follows: any new technology will eventually be forgotten if it is not considered significant in some way. In order for it to be materially, economically, legally, politically, socially, and culturally approved beyond its original location and scope, it must of course work effectively in those new locations. But to arrive there, it first needs to be exhibited, promoted, linked to policy programmes and, finally, seen as useful by investors, producers, and users. From their invention, new technologies are rooted in technological networks and need to be compatible with the standards of these networks. Technological innovations are always caught in a dilemma between the need to fit into a given technological network so as to be recognised for what they promise, and the conflicting need to promise something truly new that changes the status quo of how things have been done. The complex evaluations and judgements that play a crucial role in living with this dilemma would be misunderstood if it were assumed that innovation could be explained solely in terms of the superiority of a technology that easily travels.

To accentuate this important point, let us rephrase it. Technological innovation cannot be explained either by treating it as a rational economic choice or by criteria of greater technological efficacy. This is because such calculations depend on larger material, economic, political, legal, technoscientific, social, cultural, and aesthetic frameworks that inevitably defy explanation by rational

---

9   An illuminating example can be found in Uli Beisel's article on mosquito nets (2015). In another article, Beisel, alongside her two co-authors, demonstrates the importance of the way in which a new technology becomes entangled with existing networks of technologies and institutions (Beisel et al. 2016).

choice. Rational choice only works within these frameworks and not above them (DiMaggio and Powell 1983; contra Rogers 1995). To be valued by relevant publics, new technologies must also reflect existing normative imaginaries of the future that are captured in narratives. Both the larger historical frameworks and the narratives of past imaginaries of the future leave traces that are stored not only in the cultural memory, but perhaps more reliably in various museums and libraries that collect specimens of all kinds, old technologies inscribed with related procedures, science, literature, works of art, films, and audio recordings—most explicitly science fiction books and films[10] Across these repositories, we can thus identify a "technological archive" from which the registers of translation are drawn.

To explain the difference between valued and less valued technologies, Werner Rammert (1999, 33) proposes that all the elements and processes that come to play a role in the evaluation of a technology belong to what he refers to as a "technological archive". The fundamental argument is that the criteria for evaluating a new technology cannot be completely new and must therefore be drawn, at least in part, from existing registers. These registers are stored in the technological archive, and while some of them have a greater presence in the collective understanding, others need to be excavated from deeper layers of sedimentation. It is these registers that facilitate the translation of travelling technologies.

While affirming this argument, we would like to add that the boundaries between different technological archives are highly permeable and do not correspond in all respects to the territories of countries, continents, religions, languages, or fields of expertise and practice, but rather cut across them (Mumford [1934] 1963). We argue that archives form their own zones with extensions that depend on the history of the circulation of technologies. The extensions of these "technological zones," as Andrew Barry (2006, 2013) refers to them, change as the spread and scale of each technology evolves. Some universal foundations (such as the use of gravity, fire, the wheel, techniques of breeding, gathering, hunting, fishing, and agriculture) are common to all technological archives. However, other elements are found in some archives and not in others, and this distribution also varies according to the particular age and sophistication of the technology in question. We would also critically add that while it is true that valuation registers are needed to establish the worth of

---

10   Among the first and most popular fiction books related to the promises of modernity are: Capek [(1920] 2024); Orwell (1945); Jünger ([1957] 1961); and Lem ([1961] 1970). Among the first most popular non-fiction contributions are: Cassirer ([1930] 2012); Benjamin ([1936] 1969); Kracauer (1960). And among the contemporary scholarly reflections relevant here are: Jasanoff and Kim (2015); Czarniawska and Joerges (2020).

a technology (as opposed to the market value), no technology ever refers to only one register. This means that the determination of worth is the temporary result of volatile and ongoing negotiations between several conflicting registers (Boltanski and Thévenot [1991] 2006). All the chapters in this volume pay particular attention to this negotiation.

With this understanding of the technological archive as the basis for regimes of translation and valuation of circulating technological objects and procedures, we can now elaborate on an aspect of the main argument briefly introduced above. As noted, we intend to relate colonisation, understood as the subjugation of larger parts of the world by European imperialism (leaving out older instances of colonisation), to a more general definition of the term which refers to the colonisation of the lifeworld by technology. To find out exactly how they are related, we first need to look more closely at the analytical concepts of "lifeworld" and "technicisation" in order to outline the argument about the colonisation of the lifeworld. With these questions, we now enter the second hermeneutic circle of our inquiry.

4   Technicisation

4.1   *Lifeworld and Technology*
The concept of "lifeworld" emerged in the context of phenomenological debates and has had many different trajectories. We are particularly interested in the line that goes back to Edmund Husserl ([1934–37/1954] 1970; Blumenberg 2010).[11] Although, as elaborated above, an idea always has many sources,

---

11   Our approach was initially inspired by Bernward Joerges' "Technik im Alltag" (1988). Joerges and the authors of his collection focus on different aspects of technology in everyday life. We mainly develop one of them (relief of sense making through technicisation) and add a new one that was not part of Joerges' problematisation. One of the authors in Joerges' collection, Werner Rammert, drew on Hans Blumenberg's phenomenology of technology and developed this approach a few years later in an independent publication (Rammert 1999). Rammert's task was to draw up a genealogy of authors who changed the view of technology from an essentialising to a relational one. While we are inspired by Joerges' problematisation and Rammert's point about relationality, we want to make a different and new point. In 2013, the journal *Zeitschrift für Kulturphilosophie* published a special issue on technics (Konersmann and Westerkamp 2013), in which the article by Christoph Hubig (2013) situates Blumenberg's phenomenology of technology within the long and complex debate on technology and the lifeworld. However, we believe that Hubig does not fully acknowledge how Blumenberg explains the self-correcting mechanisms of technology (Hubig 2013, 264). We focus on Blumenberg's notion of the "suspension of meaning" (*Sinnverzicht*) to explore new forms of meaning-making (*Sinnbildung*) that can emerge when technology travels.

one has to start somewhere, and in this case, it is appropriate to start from Husserl. In his tradition and in a simplified summary, the concept of lifeworld assumes that, for a community of subjects, their common lifeworld refers first and foremost to the meanings that constitute their common language, and thus implies an ordered arrangement of significations that are related to practices and materialities that stabilise each other. This way, the concept refers to a given and always already understood reality: an intuitive and immediate resource of self-evident and no-longer and not-yet questioned knowledge that grasps what people do when they do what seems natural to them. This knowledge is no-longer fully understood because its meaning has been suspended by routinised use, and it is not-yet understood because its use has not-yet been disrupted by changes in the world that will inevitably arise. The important point that distinguishes this genealogy upon which the notion of lifeworld is based from others is that technology is understood here as situated within the lifeworld and not as its antagonist.

The principal method developed by Husserl to work with this analytical concept is known as "phenomenological reduction". The application of this method means that every assertion is "to be suspended that could not be traced back to the immediate givenness of consciousness—above all, the 'general thesis' of the transcendent existence of a world independent of consciousness" (quoted from Blumenberg [1963] 2020, 376). Phenomenological reduction thus requires that the analyst of the world refrains methodologically from referring to the transcendent existence of the world, as if it were a kind of corrective available to the analyst.

The phenomenological reduction became influential in sociology through the work of Alfred Schütz and Thomas Luckmann (Schütz [1932] 1967; Schütz and Luckmann 1974), where it was used to establish the premise that people do not live in the world but in what they believe the world to be, i.e. that people live in their lifeworld. After Schütz and Luckmann, the phenomenological analysis of society circulated widely, was critiqued, radicalised, and refined; it was forgotten and reinvented many times, and during the last few decades it has again been influential in several versions, some of which are close to pragmatism and to hermeneutics (see for example Moran 2000; Crapanzano 2004; Jackson 2013; Ihde 1979, 1990).

For our purposes here, the direction given to phenomenology by Jürgen Habermas in his *Theory of Communicative Action. Lifeworld and System* ([1981] 1989), helps to clarify the point in question. This book can be seen as the most authoritative formulation of a particular juxtaposition of system and lifeworld. It captures the widespread sense that the desirable, authentic, self-directed, self-sufficient, and colourful lifeworlds that exist universally are threatened by alienating and subjugating forces such as markets and technologies. Habermas who, like Blumenberg, writes after the Nazi technicisation of mass murder and genocide, amplifies Husserl's concern and identifies technicisation as the central pathology of modern society. When he speaks of the "colonisation of the lifeworld," he argues that the latter is characterised by value-rationality and reason, and progressively becomes overruled by technocratic systems of domination, which methodologically follow a means-end rationality. In a different vocabulary, we would say that the technocratic system exerts an epistemic violence on the mundane knowledge of the world. The exploration of technicisation in relation to the lifeworld from this particular vantage point in Habermas' writings (and those before him including Horkheimer 1947 and Marcuse 1964), entailed a concern with and drew attention to the dangers of technology (Feenberg 1996, 1999; Boyer 2005).

The terminology offered by this interpretative scheme is similar to one's common sense. We can all easily convince ourselves of this, for example, by considering how digital technology has colonised the workings of bureaucracy, work life, and private life. This way of looking at things is seductive, spreads easily, and invites resistance to technicisation. With the collection of case studies in this book, we argue that the "colonisation of the lifeworld" trope persists because its critique has not gone far enough. It has not shown with sufficient clarity that the lifeworld inevitably has a technical side and, conversely, that technology cannot function or even exist without being part of the lifeworld.

In what follows, we will outline a particular critical turn to the longstanding question of technology. We will borrow analytical concepts from Hans Blumenberg's studies of technology, especially as elaborated in his seminal essay *Lebenswelt und Technisierung unter Aspekten der Phänomenologie* (Blumenberg 1963). By critically following Husserl, Blumenberg asserted that technology cannot be conceptualised as existing apart from and in opposition to the lifeworld. He held that technology must rather be recognised as a constitutive part of the lifeworld—as are routinisation, formalisation, standardisation, and rationalisation together with other forms of meaning-making. Given the relevant contemporary debates, it is helpful to begin our elaboration of this position by outlining two more familiar positions on technology and the lifeworld before delving more deeply into Blumenberg's thoughts.

First, there is the widely used notion of "agencement" (translated into English as "assemblage") coined by Gilles Deleuze and Félix Guattari ([1980] 1987), without reference to Blumenberg who wrote about the idea twenty years earlier. In their understanding, the capacity to act, namely agency, is distributed among the elements of an assemblage which constitute each other. This is very much the point that Blumenberg makes, but without subscribing to

Deleuze and Guattari's other central tenet, namely that all the elements play symmetrical roles.[12] In Blumenberg's analysis, the human contribution to the assemblage stands out because it involves a unique capacity for reflexivity. Reflexivity is understood as the ability of individuals and collectives to apply the forms of explanation used for the actions of others to themselves and to orient their own actions accordingly. The assumption that human beings are responsible for their own actions and can hold each other accountable is based on the capacity for reflexivity—this point will emerge centrally when we come to consider decolonial technicisation. It is important to note that Blumenberg nevertheless maintains a relational ontology: the elements of the assemblage are co-constitutive of one another.

Second, most contemporary positions in STS inquire into routine technoscientific practices by paying attention to how they unfold in the everyday life—inside and outside of laboratories.[13] In STS, practice is usually understood to involve human action and material things inscribed with meanings and rules. So, we might say that technicisation involves people, hardware, and software. It intertwines material things (e.g. pipes, wires, gears, computer silicon chips) with rules (e.g. algorithms) and people who have embodied knowledge about the artefacts and how they work. Praxis involves the habitualisation of bodies, the mechanisation of procedures, and the algorithmicisation of signs (Rammert 1999, 35ff).[14] Over the past fifty years, STS has consistently demonstrated the validity of this approach and the insights it can provide. It asserts that technoscience is not situated outside material, economic, political, legal, social, and cultural processes, and that technology does not necessarily lead to social pathologies, nor does it disenchant the world or ruin the aura of human creations.[15]

An important consequence of the assumption that technoscience is not outside of society and its web of beliefs is that the practices of technoscientific inquiry change not only the objects, their relations, and thus also the

---

12   In STS, the relevant understanding of symmetry was introduced by David Bloor ([1976] 1991) and meant that plausible and implausible propositions must be analysed in the same way. This stance was redefined by Bruno Latour (1991) to mean that human and non-human elements of an assemblage play symmetrical roles. The exact meaning of this position remains somewhat vague and controversial.

13   See for instance, Pinch and Bijker (1987); Joerges (1988); Wynne (1988); Latour (1992); Oudshoorn and Pinch (2003).

14   The notion of "body techniques" goes back to Marcel Mauss ([1934] 1950). A more recent, comprehensive, and succinct phenomenology of the human–technology relation has been published by Don Ihde (1990).

15   For a fine example see Hennion and Latour (1996).

human users of those objects, but ultimately also the intentions and per-ceptions of the scholars who do the inquiry. From this perspective, what is often taken to be a satisfactory explanation of processes—namely, the iden-tification of an original means-end logic behind them—appears implausible. Instead, the same processes are seen to consist of an endless series of recip-rocal step-by-step adaptations. In a given practice, human intentions shape objects and artefacts, but at the same time also adapt to the resistance of the objects and the demands of the artefacts. Recognising this gradual process of mutual adaptation fundamentally changes the understanding of the structure and ontology of the relationship (we will return to this point later when we discuss Heidegger's contribution to it). Accordingly, STS research focusing on the investigation of this relationality cannot step out of it—it is unavoidably part of it (Luhmann [1990]1994; Pickering 1995). In this sense, even without engaging Blumenberg, STS scholarship avoids any juxtaposition of technosci-ence (operating with a means-end logic) and lifeworld (operating with a sense-making logic) as two separate fields.

In the previous paragraphs we placed the question of technology and the lifeworld in the context of approaches that we assume will be familiar to many readers. We have done this in order to provide an accessible entry point for our argument and to mark where our essay intends to contribute to STS. We hope to identify the conditions of possibility for doing decolonial translations of travelling technologies. With this in mind, we argue with Blumenberg that technicisation is an aspect of the lifeworld insofar as the lifeworld is not only about meaning-making, but also about renouncing meaning. Let us begin by adopting Blumenberg's summary of Husserl's thesis, which started the debate:

> Technization is the 'transformation of a formation of meaning which was originally vital' into method, which then can be passed on without carrying along the 'meaning of [its] primal establishment' that has shed its 'development of meaning' and does not want to acknowledge it any longer in the sufficiency of mere function. (Blumenberg [1963] 2020, 381).[16]

In other words, technology is one of several ways of achieving everydayness. It facilitates routinisation which, in turn, allows most mundane activities to

---

16   The translators of the collection of essays by Hans Blumenberg, namely Hannes Bajohr, Florian Fuchs, and Joe Paul Kroll published in 2020 by Cornell University Press rendered the German "Technisierung" into "technization" (see Blumenberg [1963] 2020). We have decided to use the more common translation "technicisation".

be taken for granted and to remain unquestioned. Technology aims to replace cumbersome and deeply unpredictable human effort with more predictable and standardised procedures. The result is a lifeworld stabilised by sociotechnical equipment. The notion of technicisation thus avoids pitting the mechanisms that drive technology against those that drive lifeworlds. It emphasises that we are dealing with an open-ended process of eternal entanglement, becoming, and metamorphosis.[17]

Technology, then, is not a term that denotes a particular essence. Rather, it refers to a particular way of worldmaking, both in terms of constructing equipment to secure and optimise practical attempts to inhabit the world, and in terms of reflecting on the implications and malleability of that equipment. In this understanding, an "antinomy" between optimisation and reflection is the central characteristic of technology. This antinomy makes the danger commonly ascribed to technology—namely, that it will prioritise performance over meaning, and ultimately extinguish the meaning of life—understandable, and at the same time opens up a way of dealing with it.[18] Conceived in this way, "lifeworld as technicisation" opens up a new perspective on colonisation, both as the historical subordination of large parts of the world by a few European countries and as the subordination of the lifeworld. In pursuing this hypothesis, our argument differs not only from what the Frankfurt School has taught us, but also from what STS has taught us.

So far, we have defined alongside Blumenberg how we understand technicisation as part of the lifeworld. While this position does not deny that technicisation is entangled with material, economic, political, legal, cultural and aesthetic processes, it asserts that technicisation is not determined by any of these, and conversely does not determine any of them. In the following steps, we seek to develop Blumenberg's proposal by bringing it more in line with his own understanding of technology as *relational*. We will further differentiate

---

17   While Blumenberg does not pay attention to specific realms of the lifeworld, and does not, for instance, elaborate on the mundane life of organisations with their specific workplace routines, we would like to add that the latter—contrary to common understanding—are outstandingly characterised by technicised lifeworlds. This point has been strongly emphasised in organisation studies (Whyte 1948; Strauss 1978; Czarniawska-Joerges 1992; Joerges and Czarniawska 1998; see also most of the chapters in this volume) as well as in STS (see above).

18   Hubig (2013, 266–69) argues, with reference to Hegel and hermeneutics, that Blumenberg's phenomenological analysis of technology is (necessarily) limited by its inability to reflect on its own terminology. While we see his point, we argue that this very ambitious critique need not prevent us from using Blumenberg's analysis to help us empirically identify where exactly the space for decolonising technology is to be found.

what he calls the "antinomy of technicisation" by focusing more than he does on praxis and its empirical variations. In the following two sections, still within our second hermeneutic circle, we will pursue these two critical differentiations—relationality and the praxis of technicisation—by looking more closely at the question of sensemaking and its suspension. After these elaborations, we can return to the analytical vocabulary derived from the verb "to colonise" and make our point.

4.2 Lifeworld as Technicisation

As outlined above, at the heart of this collection is the ambition to combine two lines of inquiry. First, together with the contributors to the collection, we explore empirical case studies of circulating technologies in African contexts. Second, we critically explore Blumenberg's understanding of lifeworld as technicisation. By combining these lines of inquiry, we explore the plausibility of Blumenberg's analysis and whether it can bring a new perspective to the case studies and to our understanding of the global circulation of technology and the resulting technicisation in general.

According to our understanding developed so far, lifeworld is about routinisation and its most effective form is technicisation. At the heart of technicisation is both its capacity to increase the effectiveness of means-end schemata and its ability to simultaneously suspend the meaning of these schemata (Blumenberg [1963] 2020, 391). With these two capabilities, it lends itself to being intertwined with any form of organisation, including those seeking to produce goods and services more profitably.

Technicisation relieves one of the burdens of reflecting on the full meaning of what one is actually doing while engaging in a routine practice such as a scientific method, operating a machine, doing bureaucracy, or running everyday errands. Relieving oneself of the burden of sensemaking is perceived as a beneficial effect that enables successful routinisation to happen. This is the particular twist that Blumenberg gives to a terminology introduced by Husserl: Where Husserl was worried about loss a of meaning (*Sinnverlust*), Blumenberg proposed to rather see a suspension of meaning (*Sinnverzicht*). In short, suspension or renunciation or deferral of meaning-making is a necessary element of technicisation. As Kaerlein (2013, 657) puts it, "one cannot start from scratch in each attempt to solve a new problem". Relying on a routine also means trusting it, being unconcerned even if one does not fully understand how it works or why it came into being in the first place. While Blumenberg argues that suspension of meaning through technicisation need not be feared, at the same time he does not question Husserl's diagnosis, that technicisation also causes a temporary loss of meaning (Blumenberg [1963] 2020, 381). He argues that

technicisation involves an antinomy: it cannot produce suspension of meaning without simultaneously causing loss of meaning. For him, however, the latter is the presupposition for the construction of new meanings (*Sinnbildung*).
Developing this argument further, we will show how repositories stored in
technological archives can help to identify a reconciliation between loss and
suspension of meaning.

4.3   *Technicised Sensemaking*

In Blumenberg's view, technicisation begins with an original insight and, as it
becomes routinised, trades an increase in efficacy for a loss of insight. You end
up with the ability to do something, but without necessarily having insight
into what it is that you are doing. At the beginning there has to be someone
who starts to do something technically and is aware of why they are doing
it and what it means and does in the bigger picture of things (Blumenberg
[1963] 2020, 364). While technicisation deliberately hides the original insight
from which it could only be born, the latter is preserved as a sedimentation
of insight in the technological archive from where it can be reactivated when
needed.[19] In the process, something of the past (original insight) is hidden
and something of the present (performance) is pushed towards a future that
promises infinite increases in performance (Blumenberg 1986). The process is
designed for the infinite, which it can never reach, and is in danger of losing
touch with the finite capacity of the human being in terms of time, space, and
knowledge (see Park, in this volume).

The important question for our argument is what happens to this design
when technologies travel in space rather than time? They usually travel with
some parts of the organisational form to which they belong, and with some of
the relevant skills (know-how) to use them. But they cannot take with them
the full knowledge of what they are and to which larger scheme of things they
belong (insight). This, we will now argue, constitutes a challenge and at the
same time opens a space for changes and innovations guided by praxis. Having
introduced the notion of creative adaptation as an innovation earlier in the

---

19   Husserl, in his inquiry into *The Crisis of European Sciences* ([1934–37/1954] 1970), makes
      this point by looking at the history of geometry. He argues that while ancient geometry
      was "conscious of its origin in the idealisation of the physical world," it was forgotten in
      the modern age so that a purely technical handling of the inherited art became prevalent
      (Blumenberg [1963] 2020, 380). By focusing on technicisation in the history of astronomi-
      cal knowledge, Blumenberg tells a different and less fearful story (Blumenberg 1975; see
      also Ihde 2016). It is worth noting that Blumenberg's philosophy of technology is part
      of his larger analysis of modernity as outlined in his seminal book *The Legitimacy of the
      Modern Age* ([1966]1983) that has been written contra Heidegger.

text, we can now get to the heart of the matter in just a few steps. Let us first add a few words about Heidegger's investigations into technicisation for the sake of a clarifying contrast.

In *Being and Time* ([1927] 2009) Heidegger coined the concept "readiness-to-hand of tools" (*Zuhandenheit*) with the intention to underscore that technology cannot be understood as a set of tools toward which we direct our cognition as something out there. This, he argues, is because technology and cognition are inextricably fused. We must therefore consider that both, technology and cognition, are embedded and entangled in praxis and in our own bodies, which are part of that praxis. In this vein, technology is defined as a mode of ordering that creates a specific way of being which affects how people live, who they are, what they fear, what they value, and how they imagine the future.

More specifically, Heidegger argues that "revealing objects" (*Entbergen*) in the modern world requires their "enframing" through "equipment" (*Gestell*). While enframing reveals objects, it conceals being (*Verbergen des Seins*) (Heidegger [1962] 2010, 103–05). Accordingly, the technological mode of being is claimed to orient all human striving toward effectively and efficiently achieving ends depleted of their being. The use of modern technology with its instrumental logic—its means-end schemata—so the argument goes, conceives phenomena such as nature, plants, humans, and animals as mere resources (or means), waiting to be exploited for particular ends. For example, "wood is a forest of timber, the mountain is a quarry of rock, the river is water-power, the wind is wind 'in the sails'" (this being an argument Heidegger already made earlier; cf. Heidegger [1927] 2009, 100). Alongside Blumenberg and Heidegger, we support the general observation that equipment is not a set of passive objects at our disposal. Rather, equipment is the material side of a mode of enframing that shapes a particular way of being and, with it, a way of thinking. What we are arguing against Heidegger, but following Blumenberg, is that technology necessarily conceals being, and that technicisation is therefore a form of pathology. In order to sharpen this point and make it useful for the next step, we re-iterate some of the argumentative steps we have already taken.

The full sense and meaning of a sociotechnical process is deliberately delegated to (or inscribed into) a technical arrangement. It is concealed, or thinned out, so that users can concentrate more on functionality and efficacy. After an initial human trigger, this arrangement can operate in an almost machine-like manner without much further human intervention—in Heideggerian terminology, it thus becomes equipment (*Gestell*). However, a less essentialising and ontologising and more praxis-oriented approach reminds us that this concealment can hardly be permanent. A routine operation delegated to a machine will not be trusted forever. Rather, it works like the embodiment of knowledge through the training of bodily techniques, like learning to write with one of your hands. What your writing hand knows better than your mind will be painfully brought back to your mind if you injure your writing hand and have to learn to write with the other hand.[20]

Being comfortable with suspending sensemaking and trusting routines is something that lasts only as long as the routines are—at least for some people—going well. The point is that the people who benefit from the routines can be unconcerned. They can even forget why the routines work well and forget that they work less well for others. As soon as some routines begin to go wrong for the same or other people—or achieve something that was not expected (see Schritt, in this volume)—the search for meaning returns. Concealment of meaning through technicisation and its occasional exposure constitute each other as two sides of the same coin. This also implies that the arbitrariness, contingency, lack of certainty, and even the injustices that are always inherent in technicisation disappear in the hinterland. How long specific sociotechnical assemblages hold together and are trusted, depends largely on two things. It depends on the effectiveness and plausibility of their original translations of sensemaking into a method that serves to conceal meaning and hide the method's lack of certainty. And the durability of sociotechnical assemblages also depends on the extent to which they are accepted and trusted also by those who do not actually benefit from them. This, in turn, implies that without more or less radical disruptions, technological innovation would be less likely to occur. With this observation in mind, we argue, that the seemingly plausible fear that technicisation leads to an erosion of the meaning of human existence—or, in Husserl's prophetic words, that technology prompts us to "take for true being what is actually a method" (Husserl quoted in Blumenberg [1963] 2020, 382)—is occasionally and under certain circumstances justified, though not as a general principle and not as a teleological trajectory of history.

The many versions of the Golem legend are the most popular and widespread expression of the deep fear that technology will outmanoeuvre human control. It was this fear that inspired the high theories of technology that emerged in the early 20th Century in the rapidly technicising countries of Europe. Now, in the early 21st Century, it is being reactivated in many, but not

---

20   For broader arguments in support of this assertion see Marcel Mauss on *body techniques* as forms of memory (Mauss [1935] 1950). But importantly, see also Maurice Halbwachs ([1939] 1991) on *collective memory* and Jan Assmann (1992) on *cultural memory*, as both argue that memory entails material things and technics. While Assmann primarily thinks of writing as the key cultural technic, his argumentation would in fact gain by giving other technics a similar importance. See also Joerges (1996) on technology as the body of society and Ihde (1990) on embodiment and memory.

all, parts of the world. It reappears as the spectre of our time. In the face of doomsday scenarios inspired by global warming, huge demographic shifts, and related struggles, new and incredibly powerful Golems are emerging. They promise means of averting doom at the last moment, while at the same time creating dramatic fears. Two in particular stand out: genome editing and artificial intelligence. We argue that this fear prevents a more accurate analysis of the main problem. We therefore attempt to differentiate its core aspect—the deferral of sensemaking—by making the analysis outlined so far more pragmatic and sociological. In doing so, we seek to provide new insights into how the deferral of sensemaking, disruption, and creative adaptation relate to each other, and to the circulation and translation of technology.

Borrowing Blumenberg's notion of technicisation, we are interested in the empirical variations produced by the ways in which technology is translated and made to travel, and what this means for the contemporary. Sensemaking—including the suspension of it—is not simply a presupposition of praxis, but at the same time, a result of it. Heidegger acknowledges that technology ontologically precedes science and sensemaking when he speaks of "readiness-to-hand of tools" (*Zuhandenheit*), as we explained above. Sense is an achievement of praxis that cannot be fixed forever after it has been attained. Instead, it requires repetition over and over again and in this process it inevitably changes.

To gain empirical access to the workings of these changes, one can examine different situations and events of the same kind of technicisation, or different kinds of technicisation in the same situation and event. In practical terms, this can mean paying attention to the circulation of concrete technical objects, for example, exploring the workings of mobile phone chain texting (Schritt, in this volume), the breathalyser (Biecker, Kagoro and Schlichte, in this volume), of the automated public water dispenser (Tristl and Boeckler, in this volume), or of the formal organisation of access to medical treatment (Park, in this volume) as they travel from one context to another. For obvious reasons, technical solutions tend to emerge first in communities of practice that face a particular challenge, but also have relevant technological archives and infrastructures, webs of beliefs, and networks of institutions that facilitate their emergence and integration. However, when technologies travel and begin to circulate in other contexts, several things change. This is already the case when a technology leaves the laboratory or workshop and is put into regular use. It often turns out to work differently from what was expected in the laboratory and, as a consequence, requires adjustments (Wynne 1988).

A similar process is triggered when a technology moves from one context of use to another. A series of translations and adaptations are required to make the new technology work. This process is mostly accompanied by disruptions to the taken-for-granted sociotechnical assemblages that operate in the new context. These disruptions may be more or less accidental, imposed or desired, more or less beneficial or harmful. Even when they are intended, they never unfold exactly as planned or intended by anyone involved. Contrary to the promise of perpetual progress that accompanied modernity's technology, interventions in complex assemblages cannot possibly only have intended consequences. Most often the unintended and unpredictable consequences actually dominate, causing further disruptions. Our main point here is that these inevitable contingencies and disruptions, while creating challenges, open up spaces for creative adaptation and innovation. Losing trust in previously taken for granted technicised routines becomes a necessary condition for innovation (see for this point Arendt [1958] 1998).

On the basis of our considerations so far, we can now claim that a critique that accuses technicisation of colonising the lifeworld is based on a false assumption. The lifeworld does not offer itself as a superior standpoint, as a foundation for a truly human, unadulterated, authentic knowledge in harmony with nature, free from conflicts of interest, greed, ruthlessness and evil. A seemingly immaculate lifeworld does not provide an Archimedean point for the lever of the critique of technology because it simply is not immaculate and does not exist outside of technology. More generally, if being human and using technology are co-constitutive, then human beings cannot be conceived as ontogenetically and phylogenetically preceding the evolution of technology. From this perspective, technology is a mode of ordering that creates a specific way of being which transforms how people live, who they are, what they fear, what they value, and how they imagine the future. Accordingly, this introductory essay seeks to identify an alternative position from which a critique of technology can be expressed from within sociotechnical assemblages. This requires a final step within our second hermeneutic circle before we can enter the third one.

4.4   *Technicisation and Decolonisation*

There is a considerable and still rapidly growing number of publications that seek a clearer and more differentiated terminology to better understand the process commonly referred to as "decolonisation". Although these efforts have been going on for several decades, there is little consensus on the correct concepts and theoretical approaches. Rather, the struggle for a better approach continues, the debates are highly politically and emotionally charged, and the importance of positionality is often overstated. With this essay we do not aim to identify the main trends and provide an overview of the ongoing heated debates (cf. Lyons, Parreñas and Tamarkin 2017), nor do we aim to invalidate other contributions (cf. Tuck and Yang 2012). While we build on and strengthen some of the arguments in this literature, we are certainly a critical provocation to others. However, we believe that the many issues involved in decolonisation can and should be approached in a variety of ways. In this sense, we offer a fresh look at the matter, one that brings to light several issues that are either not seen or not given sufficient weight by most other approaches.

In the introduction to this essay and again in the context of the first her-meneutic circle, we pointed out that the verb "colonise" as used in the phrase "technology colonises the lifeworld" obviously has a different meaning from the way it is used in the phrase "Europe colonised Africa". Accordingly, "decolo-nise" seems to refer to two different political agendas. The first focuses on pre-venting technicisation from continuing unchanged and on restoring what has been damaged or lost by technicisation wherever it has occurred. The second agenda is to restore the legal and political sovereignty of formerly colonised peoples so that they can live in self-determination.

For most people living in Europe and former European colonies, the word "decolonisation" originally referred to gaining political independence from colonial powers. Gaining sovereignty over one's own affairs was a decades-long and often violent process that was achieved gradually, with some milestones in 1947 (India), the 1960s (most African countries), 1991 (most former Soviet republics) and 1994 (South Africa). Decolonisation as the main policy goal was part of a broader understanding of the modern state, with bureaucracy and related technology at its core, underpinned by the ideology of the nation as the bearer of the state. The new sovereign governments had little doubt that their mission was to lead their countries to progress, and that this could only be achieved through technoscientific modernisation. Postcolonial governments and elites also did not question nationalism as a new form of genophilia that filled and justified the state as its shell, but rather understood it as a form of decolonisation that would help overcome tribalism. Moreover, this policy saw itself as part of a universal trend to use development through technological progress to strengthen the state, after the Soviet Union, Japan, China, and India had become, or were about to become, leading centres of technology and its enormous profitability.

In this spirit, post-independence African political elites and their govern-ments embarked on a technoscientific development trajectory that had largely been denied to them by colonial governments. With the technocratic elite, a new and potentially powerful social class emerged to drive this agenda. In the same early post-independence period, the intellectual elite was more concerned with the decolonisation of the mind (Thiong'o [1981] 2005).[21] In the first enthusiastic years of independence, many intellectuals in the humanities understood technoscientific progress as a process that could be left to engineers, scientists, and technocrats. Some saw it more as a tool for decolonisation, while others saw it as a Trojan Horse, bringing new technologies that would soon create new dependencies. Léopold Sédar Senghor was perhaps the most prominent *l'homme de lettres* to see technology as a tool for decolonisation (UNESCO 1974). Ghana's first president, Nkwame Nkrumah, was also initially a strong supporter of this idea, which materialised in the Akosombo Dam with its hydroelectric plant on the Volta River (Miescher 2022), and in Ghana's ambitions to develop non-military uses of nuclear power since the 1960s (Osseo-Asare 2019). Later, both grand projects of the newly independent Ghana also came to symbolise the pitfalls of the modernisation they represented.

Both technocratic and culturalist approaches to decolonisation were soon criticised for failing to see the other important forces at work. Samir Amin (1988) famously criticised both in one fell swoop. Culturalist approaches obsessed with the decolonisation of the mind would obscure the real workings of capitalist colonisation. In particular, he attacked Edward Said ([1978] 2003) for deriving his entire argument from the assumption that one must first have a liberated mind and then all the rest will follow. Technocratic approaches, Amin's argument implied, would similarly obscure the market forces driving the technology-based development interventions of the new independent governments. Shortly before Samir Amin, Paulin Hountondji ([1983] 1990) argued that scientific and technological activity, as practised in Africa in the 1980s, was as "extroverted" (i.e. externally oriented or outward-looking) as economic activity on the continent. He saw this as the result of the subordination of African knowledge systems to the world knowledge system—a mechanism that he believed worked in a similar way to the integration of subsistence economies into the world capitalist market. According to Hountondji ([1983] 1990, 7), both had led to an underdevelopment that was not the result of an original backwardness but its cause.

During the 1980s it became hard to ignore the fact that the attempts to accelerate technicisation on the African continent were largely shaped by

---

21   This is not to say that the concern with decolonising the mind faded, or should have faded, in later years. Some years later, looking at the 1980s, Achille Mbembe (1992) brilliantly examined the interconnectedness of the damaging external political and economic developments mentioned above with troubling internal cultural and political developments in Cameroon. He unearthed the deeper layers of what was emerging as a new political form, the "postcolony", by focusing on the role of the grotesque and obscene in struggles for the legitimacy of power in Cameroonian politics (Mbembe 1992).

globalised and financialised markets, by the interests of global powers, the mechanisms of the Cold War that since 1947 had driven countries to align themselves with one of the two opposing political blocs, and ongoing forms of epistemic inequity (Osseo-Asare 2019). Against this background, it seems plausible that accelerated technicisation in Africa would go hand in hand with forms of extroverted orientation and resource extraction, accompanied by forms of extraversion (Bayart [1989] 1993). It therefore also seemed plausible to view imported technologies and scientific innovations with suspicion (Mudimbe [1982] 2023, 39–40; Hountondji [1983] 1990, 7; Mavhunga 2017, 1).

Even if they were not clearly separated in practice, at the level of discourse and political rhetoric two seemingly opposite perspectives continued to be deployed: one emphasized independence through the transfer of cutting-edge technoscience, the other warned that this was set to be another Trojan Horse. From the theoretical standpoint developed in this essay, it is important to look at what lies behind the continuity of this political rhetoric to the present day.

As we have argued, in practice you cannot import a single technology (such as drones) or machine (such as a nuclear magnetic resonance spectrometer) because it has to be translated to work in the context of another set of technologies, which in turn only work as part of an even larger set of networked technologies that include their material, technical, scientific, economic, political, and legal infrastructures, all situated within a sufficiently uncontested web of beliefs that makes the whole enterprise plausible and trustworthy. A tricky challenge, arising from the fact that a technology can only work in a network of other technologies, is where to start promoting technicisation—for example, if there is a city suffering from a shortage of drinking water, sewage problems, and periodic outbreaks of cholera (Rottenburg [2002] 2009). So, trying to keep a particular machine or technology out of a particular technological zone because it is seen as harmful or useless, and importing another because it is seen as useful, can only work to a very limited extent. In practice, therefore, the question is usually how much of one technology, and what translation of it to choose in order to have some of the other.

Even taking all this into account, the relationship between the intended technicisation and the unintended material, economic, political, legal, and cultural effects is in most cases not an either-or, but rather a this-and-that structure. If we assume a binary yes-no logic here, we fail to understand the translation of travelling technologies. Instead, we have to assume that there is an endless series of step-by-step translations that rarely result in what was originally intended, and which in turn require subsequent translations that could not have been foreseen. More generally, the key question is how to create a zone of decolonial translation of circulating technology in which this assessment can be oriented towards the common good. And then, of course, the common good has to be constantly renegotiated at local, national and international levels, between these levels and between different logics of constituting the common good. This means that the introduction of a new technology—take for instance contraception or in vitro fertilization—is usually accompanied by controversy, the various positions and parties of which do not automatically follow existing groupings, but tend to change their composition.

This essay takes a fresh look at this ongoing process of muddling through, which cannot be explained by invoking political agendas promising a better future, deep cultural scripts discerned in the history of a place, or teleological theories trapped in a false binary on the question of technology. By focusing on global flows, in particular the circulation of technology, and reopening the question of modernity's technology, we ask more realistic questions: If the successful translation of circulating technologies always and everywhere changes (colonises) the routines of lifeworlds through their technicisation, does this necessarily imply material, political, economic, legal, cultural, and aesthetic colonisation? And what does this mean in particular for postcolonial countries, which largely import globally circulating technologies? To what extent are the mechanisms of colonisation caused by the different ways in which travelling technologies are translated? Who has the agency to change the translations? To whom can the responsibility for the outcome be attributed? What is the role of the technological archive in this context?

In order to find answers to these very broad questions, we have narrowed them down to a level where the empirical evidence can speak back to the theoretical assumptions. To do this, we first delineated the zone in which the translation of a travelling technology is negotiated. We were then able to identify the key question: How to achieve a desirable level of suspension of meaning (Blumenberg calls it *Sinnverzicht*), facilitated by a particular form of technicisation, without this leading to an undesirable loss of meaning (Blumenberg speaks of *Sinnverlust*)? And finally, we have been able to get to the heart of the matter and ask, how exactly can travelling technologies be translated without creating new forms of colonisation?

With this systematically derived new problematisation of a well-known and intolerable fault line, we have attempted to draw attention to more fine-grained factors that are often left out of established analyses. An important specification concerns the spaces traversed, enacted and created by global flows and, in particular, by travelling technologies. Our argument is based on a radical overcoming of methodological nationalism, which usually treats the nation or the nation-state as the fundamental unit of research and as the container of all social processes. Overcoming this approach means that regions, continents and cultural areas (such as the Orient or the Occident) cannot be understood as containers either. Instead, they too are better understood as effects of circulations, which essentially include travelling technologies. The interesting questions relate to how circulating technologies regularly transgress existing borders and boundaries and help to constitute new ones in a perpetual process that never results in the construction of a solid, everlasting container.

The importance of this methodological approach is further underlined by a second specification. If we consider that technologies also, and perhaps more importantly, flow across social fields, it becomes even clearer that these fields are themselves largely constituted by this flow. Three examples illustrate this: When the methodology of randomised clinical trials moves from medical research to economics, the boundaries between these disciplines change. When genetics offers a new way of understanding and shaping life at the molecular level, it has profound implications not only for the life sciences as a whole, but also for law, archaeology and history. When computational science and faster computers make it possible to store and work with huge amounts of data, it changes the relationship between causal and statistical explanation in all fields. The diffusion of some technologies becomes so pervasive and affects such fundamental dimensions that few social fields are left untouched. Genetics and digitalisation are perhaps the contemporary prototypes of what Nathan Rosenberg and Manuel Trajtenberg call "general purpose technology" (Rosenberg and Trajtenberg 2001).

The common point between the two specifications is that whatever aspect of space we heuristically prioritise for a particular analysis, it is never a homogeneous and fixed space. It is always a zone in constant flux, characterised by heterogeneity, incoherence, contingency, contradiction and conflict. What is also important for our argument here is that the modern nation-state, the market structures associated with it, the social stratification of the population, the different interest groups, the social classes and regions of the state territory have different expectations of and benefits from technological infrastructures. The translation of a technology in a zone of conflictual negotiation inevitably therefore results in some people benefiting more than others, and some even losing privileges they had before. This observation complicates the challenge of creating a zone of decolonial translation.

In seeking a better understanding of the term decolonisation, our essay follows Eve Tuck and K. Wayne Yang's (2012) observation that the increasingly frequent and almost arbitrary use of the term obscures rather than captures the central issues at stake. But by taking a different object of inquiry—not the return of land to indigenous peoples in former settler colonies, but the circulation of technology between different social fields and technological zones around the world—our approach is also different. Our aim is to work out in microscopic detail where and how colonisation and decolonisation take place in the practice of translating circulating technologies. To this end, we propose a number of terminological clarifications, but we have no promise of how to get past all the trouble.

It helps us to clarify our theoretical position by pointing out another way in which we differ from Tuck and Yang. It is the distinction they make (more or less implicitly) between metaphor and concept. For us, it is important to maintain that (analytic) concepts not only attempt to capture or correspond to realities, but also help to create them. This means that by using the word decolonisation not as a metaphor but as a concept, as Tuck and Yang suggest, one is not simply referring to a real practice that can be found out there in reality and taken up to be held up against the concept to see if they correspond. The general point is that any reality is the result of world-making—implying conceptual and metaphorical work—and therefore cannot be taken as sufficient proof that its conceptual representation was correct (see Davidson 1974; Rorty 1989). The suspicion that this very epistemology can itself become an instrument of colonisation is nevertheless justified. For any construction of reality is inevitably shaped by deep ontological commitments and by economic and political interests that are not inherently part of the reality one is trying to grasp (Harding 1992). However, we do not see the possibility of an epistemological shortcut to the reality of things (Lynch 2013). Instead, we see much scope for trying to find out more about how deep ontological commitments, standpoint epistemologies, and economic and political interests influence world-making and mechanisms of colonisation and decolonisation. We argue that a microscopic and praxeographic analysis of the translation of circulating technology offers a fruitful approach that invites starting from marginalised positions, wherever they may be (Harding 1992, 447f).

At the moment of translation, when the technology enters a new context, the set of material, technical, political, economic, legal, cultural and aesthetic assumptions inevitably inscribed in the travelling technology are disassembled and thus made visible. These assumptions have to be readjusted and reassembled if the translation is to be successful. It is in this moment of practice that the space to make a difference opens up more widely. In this space-time, free action becomes possible that is not completely determined by the conditions that dominate the situation. This space-time is not structured by a predetermined and fixed binary, such as that between "us" and "them". Rather, it is a zone characterised by different interests, divergent priorities, multiple conflicts, political manoeuvring, shifting alliances, strategies of concealment, facades of consensus and, last but not least, the motivation to find a compromise to keep things going. The openness of the zone allows translation to make a difference. It can either perpetuate an inherited epistemic inequality or create a new one, or it can perpetuate an existing epistemic equality or create a new one. In the latter case, we speak of deco-lonial translation.

In order to come closer to asking sharper and more empirically meaning-ful questions about this particular praxis, and thus to increase the chances of eventually finding answers to the question of technology and its relation to mechanisms of colonisation and decolonisation, we now enter our third hermeneutic circle. We will look at how the empirical case studies brought together in this book relate to the questions developed above.

5   The Case Studies

In testing the theoretical framework sketched in this essay, the chapters of this volume problematise how the translation of selected travelling technolo-gies plays out in particular contexts and situations. Each chapter focuses on how the meanings inscribed into travelling technologies and their hinterlands (or technological archives) challenge translation. The authors conclude from their empirical research that in the context of their African sites, technicisa-tion often does not result in suspension of meaning that generates smooth and taken-for-granted routinisations, but rather in various dysfunctions that keep the search for meaning alive, and even foreground it. Indeed, the dysfunc-tions have multiple and often deplorable effects. For our analysis, however, this means that it is precisely the ongoing dysfunctions that open up a space for decolonising translations. All the chapters are based on empirical research conducted as praxiographies of concrete events in concrete situations, and for all of them, the object of study is a particular technology on the precipice of being embedded in a lifeworld. Some focus more on practices and lifeworlds within formal organisations, others more on the impact that organised prac-tices have on lifeworlds outside of formal organisations. Yet all pay due atten-tion to the ways in which they are interrelated.

In their chapter, Christiane Tristl and Marc Boeckler examine the operation of a public automated water dispenser that releases twenty litres when hold-ing a prepaid card against the device's card reader. Those who designed the system wanted to increase the efficacy and efficiency of water distribution by replacing human agency at the point of sale with a supposedly incorruptible machine: you get a fixed amount of water for a fixed amount of money on your card, and everything else is irrelevant to this human-machine interaction. Yet the same process makes a desperate person in rural Kenya aware of the impor-tance and precariousness of water distribution when her prepaid water card runs out of credit. She is simply unable to negotiate her vital water needs with the vending machine as she could before when there was a human handling the same process. The case study powerfully demonstrates how the water dispenser, while attempting to defer the full meaning of access to water, occasionally brings to the fore what it was intended to conceal. The authors also show that this outcome is not only a consequence of technicisation but is in fact driven by the introduction of neoliberal reforms and market logics into the governance of sensitive public goods such as water.

Jonas van der Straeten and Jochen Monstadt, also concentrating on Kenya, explore the provision of electricity in Nairobi. They explore the engagements of a utility concerned with doing its best to keep electricity provision running quietly and reliably in the background. The case study examines the logics of this translation work and shows how it sets in motion an ongoing series of contests, negotiations, and compromises, at the heart of which are different understandings of what constitutes a public good. The promise of universal supply has been partially fulfilled in real estate projects that have built some of the wealthiest neighbourhoods in Nairobi. Here, developers set up their own companies to buy electricity from the national utility under official licence and then sell it to tenants who can pay regularly. By contract, in Nairobi's vast informal settlements, unlicensed local cartels poach electricity and sell it illegally to residents through improvised supply networks. This is accompanied by all sorts of irregularities and service interruptions that never allow the large technical system to recede into the background. The authors examine how the national electricity company, in its efforts to provide universal access to electricity, constantly negotiates compromises with various intermediaries including illegal ones. The reason for this willingness to compromise, even when the legal situation demands a different kind of intervention, is that the utility hopes to get closer to its goal of providing universal and uninterrupted access to electricity for all customers. The structural background to this particular strategy is the same neoliberal logic of privatisation and decentralisation of service provision which is explored in the previous chapter on public automated water dispensers in rural Kenya. Both studies examine attempts to turn citizens into customers who, through technicisation, are individualised, made to follow the logic of market mechanisms, and bear the full responsibility for having or not having water and electricity.

Turning to communication as another basic infrastructure of the modern state, Jannik Schritt examines mobile phone chain text messaging in the context of political mobilisations in Niger in 2011. He shows how the mobile phone became part of a distributed agency in mobilising mass protests. Focusing on the functioning and intertwining of technologically mediated and face-to-face interactions, the chapter shows the mutually constitutive nature of the two. Long before 2011, the mobile phone had become a natural part of everyday life. Its routines had become self-evident, suspending the question of what it means to live in a largely technologically connected community. Schritt shows how mass protests made the workings of mobile telephony visible once again, triggering new efforts to make sense of the role of communication technology in the realm of the political. While the process of technicisation here is challenged not by service disruptions but by a surprising success in political mobilisation, the effect is the same. More generally, the chapter demonstrates that once a particular technicisation has become a taken-for-granted part of a lifeworld, the deferral of sensemaking can also be interrupted by its surprising success. New questions are raised and new possibilities for adapting the technology to the given context emerge.

Also concerned with the study of basic infrastructures of the modern state, Sung-Joon Park's chapter examines health care and, more specifically, the expansion of mass HIV treatment programmes in Uganda between 2010 and 2014. Focusing on the temporal dimension of technicisation, the chapter engages in detail with how new technologies aimed at improving universal access to treatment are being implanted into the lifeworld of local HIV/ AIDS organisations, primarily to accelerate the roll-out of antiretroviral drugs. Park argues that his interlocutors perceive this acceleration as creating a gap between time that can be measured and quantified (in minutes, hours, days, and years) and the lived experience of time—for example, the experience of care as time spent on behalf of others. This gap is in turn experienced as a general lack of time. The technicisation of work, originally intended to speed up HIV treatment to save as many lives as possible, is inexorably widening this gap between what the author calls "project time" and "world time". The chapter shows how the widening of this gap is making the profession of pharmacy more and more a matter of knowing-how-to-do and less and less a matter of knowing-what-it-is. The author emphasizes the need for sensemaking by arguing that this drifting apart of knowing-how-to-do and knowing-what-it-is problematically erases the possibility of insight. He concludes that global health lacks not only money, resources, and know-how, but also, crucially, the "extra time" to produce the necessary insight for doing the work of caring for others. In this chapter, therefore, the inherent logic of technology to sacrifice insight for higher achievement is reinforced not by an economic calculus, as in the chapters by Tristl and Boeckler and van der Straeten and Monstadt, but by a quest for technicised acceleration to save as many lives as possible.

Staying in Uganda, Sarah Biecker, Jude Kagoro, and Klaus Schlichte focus more directly on the role of technicisation in government than any of the other five empirical chapters in the volume. The chapter specifically examines the role of files and breathalysers in the work of the Ugandan police, and the accounting procedures that donors have required the Ugandan government to follow for many years as part of their budgetary support. The authors pay particular attention to the varying degrees to which the three forms of technicisation act as instances in which a suspension of meaning becomes discernible. In so doing, they examine when and how technicisation renders domination invisible and the extent to which domination remains visible and contested in the negotiations surrounding it. The chapter shows how the logic of power reinforces the logic of technology to prioritise achievement over insight, ultimately asserting that technology helps to conceal the workings of power.

The forms and functions of technicisation of government and organisation examined in the empirical chapters above not only share a number of common and complementary aspects, but are also fundamentally interrelated at another level. They deal separately with one of the following objects of study: the provision of water, electricity, telecommunications, and medicines; the making of bureaucratic documentation, accounting, measurement, and law enforcement. However, none of these provisions and constructions can function independently of the others. Moreover, they all depend on surveillance technologies that allow all sociotechnical transactions to be traced back to the individual office holder responsible for the decision. This kind of technicised—increasingly digitalised—surveillance is seen by the interlocutors of all cases studies as promising greater accountability and transparency in the pursuit of sociopolitical order.

In this sense, Alena Thiel's chapter, the last in our collection, examines the production and use of biometric identification technologies in Ghana since the 2010s, as a process of technicisation that is officially propagated with the hope of leading to a new form of transparency. Thiel explores the consequences of the automation of identification processes—she speaks of "datafied selves"—for the construction of personhood in various social interactions. She shows how the implantation of identification technologies into the lifeworld remains contested, mainly because the technologies imply the classification of human kinds and involve haunting hegemonic categories. These create various frictions and thus become more visible when the very process of technicisation was intended to conceal them—an observation we also made with regard to the previous chapters of the volume.

6   Conclusion

The argument presented in this introductory essay, and substantiated in the empirical chapters, aims to open up a space in which the conditions of possibility for decolonial technicisation can be identified. In order to do so, we have had to circumvent, redefine, or refute a number of established dogmas. The first steps of our argument have also been pursued by other scholars with different approaches. It seemed necessary to revisit them in order to prepare the specific and new argument we wanted to make.

Our fundamental problematisation, inspired by Hans Blumenberg, was that on the one hand, modern technology has emerged as an ever-changing design that is geared towards infinity and stands in an aporetic relation to the irreversible finitude of individual and collective human existence. On the other hand, going beyond Blumenberg, we argue that the very design of modern technology is complicit in bringing the finitude of the planet uncomfortably close. We have said that the last fifty years have therefore been marked by a growing preoccupation with the antinomy between infinity and finitude, and that this preoccupation has become intertwined with decolonisation.

Against this background, we have argued that in a popular and deeply rooted understanding, technology is readily juxtaposed with the lifeworld. We were able to affirm that, as so often, two opposing interpretations of this supposed juxtaposition—technology relieving us of some of the burdens of existence versus technology depriving us of the meaning of existence—are based on the same assumptions, operating with the same simplifications, homogenisations, and essentialisations. These assumptions concern the "nature" of technology and the lifeworld, the "origin" of technoscientific innovations, the "diffusion" of technology, the definition of the relationship between territory, archive, and technology, and finally the distinction between innovation and creative adaptation.

In this sense, we rejected the ingrained idea that any one of the many forces and logics that together drive the development of a mode of existence can by itself and alone determine the course of events. Accordingly, we have affirmed the insight that technology is neither a neutral device to be used arbitrarily for good or bad purposes, nor is it good or bad *sui generis*. Technologies are relational in the sense that they cannot exist without being dependent on other technologies and infrastructures, and in addition also being dependent on materialities, laws, rules, economic resources, sociocultural beliefs, aesthetics, acceptance, and trust. And finally, we have shown

how the resulting contingency not only allows for, but necessitates, interventions for which people hold each other accountable. They do so because
there were always other routes that could have been taken. Our case studies
have shown how frameworks of government—such as the vast repertoire of
organising tools rooted in neoliberal political economy—shape the translation of travelling technologies. More importantly, they have all identified spaces that allow for different translations and creative adaptations, as
well as for multiple manipulations to navigate and often reconfigure power
asymmetries.

The claim that technicisation necessarily involves an antinomy between
achievement and insight is not new, but we have imbued it with a different
twist. We have shown how technicisation unfolds between achievement and
insight, how the implicit and irresolvable antinomy between the two effects
is folded into sociotechnical assemblages. As a consequence, we have argued
that even if we can identify some patterns in the unfolding of assemblages,
there is always a considerable degree of unpredictability. The recognition
of unpredictability is the necessary condition for the possibility of acting,
making a difference, and creating something new (cf. Arendt [1958] 1998).
Unpredictability also underlines the reasonable need to be cautious about
the inherent and insurmountable limitations of technoscientific predictions. And this, in turn, underlines the democratic need for legal accountability not only for policy decisions made with reference to technoscientific
evidence, but also for translations of travelling models and technologies of
various scales.

With this last review of the logical steps of our argument, we now turn to
a summary of our central contribution. It has been our aim to show how the
unfolding of sociotechnical assemblages is inevitably affected by the translation of travelling technologies. Once a new and promising technology has been
shown to work in one corner of the globalised world, it will soon appear in
another corner, and after a while it will be almost everywhere. But a technology that has become ubiquitous is rarely the same in each field of its circulation and hardly ever identical to the one that initially started the journey. It can
only travel, as long as it can be translated, and modified. In the process, and as
a result, a more or less different technology may emerge elsewhere, and the
same process begins all over again. To participate in technological innovation
is to participate in this work of translation.

We have argued that the quest for decolonial technicisation can potentially be realised through the translation of circulating technologies. We have
claimed that decolonial technicisation is not only oriented towards the particular and the local, but at the same time towards the universal—in the sense

first proposed by Léopold Sédar Senghor (1956), later specified by Souleymane Bachir Diagne (2013), and here referred to as the planetary. We conclude by summarising how this argument works.

The translation of travelling technologies is something unavoidable. First, as technologies become successful, accepted, and normalised, they become woven into the routines of lifeworlds. Their presuppositions, uncertainties, and fundamental premises thus disappear into the hinterland, archived and forgotten. In the terminology proposed in this essay, we say that the meaning of the technology has been suspended as it becomes a taken-for-granted part of a lifeworld. Secondly, when a successful technology begins to circulate, and most of them do, much of the suspended meaning remains at the point of departure because it is not fully inscribed in the technology but resides in the relevant technological archive. In order to become interesting for another context that does not have a similar archive, the travelling technology must be re-inscribed with meanings found in this new archive. It is this practice that we call the translation of travelling technology.

The practice is guided by regimes of translation that are part of technological archives, which in turn (among other things) distinguish one technological zone from another and thus never fully correspond to countries, continents, religions, languages, or fields of expertise and practice. The translation of circulating technoscientific models is therefore an achievement that takes place between several technological zones. With each step of translation, the travelling technology is not only transformed and appropriated to a local context, but at the same time transformed to contribute—in one way or another—to a problem of planetary dimensions. This is because translations always work in both directions, and so a new translation changes all its predecessors.

The decolonial space required for creative adaptation is an unstable and vulnerable zone of translation, easily hijacked and abused. It is not a space that can be established once and for all by a revolutionary act. The zone of decolonial translation needs to be constantly challenged, defended, recreated, and nurtured. To do this, we need a much better understanding of how the translation of travelling technologies as creative adaptation works in empirical detail. We suggest that future inquiries need to focus on how technicisation can be decoupled from the modernist quest for infinity. The main aim of this volume is to ignite debate and experimentation and to encourage many more praxiographic studies on the matter.

Acknowledgements

We have written this essay over a number of years, with some longer interruptions, and have been helped by discussions with many people. While we are responsible for any remaining shortcomings, we are indebted to the following colleagues for important inspirations. First, René Umlauf and Uli Beisel contributed to the early development of the argument. Second, the authors of this volume and their case studies have shaped our argument over the years. Third, Keith Breckenridge, Steven Feierman, and Peter Redfield provided critical comments on an early version and helped us find our focus. Fourth, a later version benefited from the comments of Bronwyn Kotzen and Faeeza Ballim. Last but not least, the two anonymous colleagues who accepted the publisher's invitation to review our book challenged us on a number of points, and we believe that the text has improved substantially as a result. The research on which this essay and all the chapters in this volume are based was carried out within the framework of the Special Priority Programme of the German Research Foundation (DFG 2019) under the project number SPP 1448 and title "Adaptation and Creativity in Africa. Technologies and Significations in the Making of Order and Disorder". This programme ran from 2011 to 2017 and Richard Rottenburg was one of its two speakers.

Bibliography

Adas, Michael. 1989. *Machines as the Measure of Men: Science, Technology, and Ideologies of Western Dominance*. Ithaca, NY, and London: Cornell University Press.

African Union. 2023. "Science, Technology and Space." Accessed December 18, 2023. https://au.int/en/directorates/science-technology-and-space.

Akrich, Madeleine. 1992. "The De-Scription of Technical Objects." In *Shaping Technology/Building Society: Studies in Sociotechnical Change*, edited by Wiebe E. Bijker and John Law, 205–25. Cambridge, MA: MIT Press.

Amin, Samir. (1988) 2010. *Eurocentrism. Modernity, Religion, and Democracy: A Critique of Eurocentrism and Culturalism*. Nairobi and New York: Monthly Review Press.

Anand, Nikhil. 2017. *Hydraulic City: Water and the Infrastructures of Citizenship in Mumbai*. Durham, NC: Duke University Press.

Anand, Nikhil, Akhil Gupta, and Hannah Appel. 2018. *The Promise of Infrastructure*. Durham, NC: Duke University Press.

Appadurai, Arjun. 1996. *Modernity at Large: Cultural Dimensions of Globalization*. Minneapolis: University of Minnesota Press.

Appel, Hannah. 2019. *The Licit Life of Capitalism: US Oil in Equatorial Guinea*. Durham, NC: Duke University Press.

Arendt, Hannah. (1958) 1998. *The Human Condition*. Chicago: University of Chicago Press.

Assmann, Jan. 1992. *Das kulturelle Gedächtnis: Schrift, Erinnerung und politische Identität in frühen Hochkulturen*. München: Beck.

Barry, Andrew. 2006. "Technological Zones." *European Journal of Social Theory* 9, no. 2: 239–53.

Barry, Andrew. 2013. *Material Politics: Disputes Along the Pipeline*. Chichester, Sussex: Wiley-Blackwell.

Bayart, Jean-François. (1989) 1993. *The State in Africa: The Politics of the Belly*. London and New York: Longman.

Bayart, Jean-François. 2000. "Africa in the World: A History of Extraversion." *African Affairs* 99, no. 395: 217–67.

Beisel, Uli. 2015. "Markets and Mutations: Mosquito Nets and the Politics of Disentanglement in Global Health." *Geoforum* 66: 146–55. https://doi.org/10.1016/j.geoforum.2015.06.013.

Beisel, Uli, Sandra Calkins, and Richard Rottenburg. 2018. "Divining, Testing, and the Problem of Accountability." *HAU: Journal of Ethnographic Theory* 8, no. 1–2: 109–13. https://doi.org/10.1086/698360.

Beisel, Uli, Rene Umlauf, Eleanor Hutchinson, and Clare I. Chandler. 2016. "The Complexities of Simple Technologies: Re-imagining the Role of Rapid Diagnostic Tests in Malaria Control Efforts." *Malaria Journal* 15: article 64. https://doi.org/10.1186/s12936-016-1083-2.

Benjamin, Walter. (1936) 1969. "Das Kunstwerk im Zeitalter seiner technischen Reproduzierbarkeit." In *Illuminationen: Ausgewählte Schriften*, edited by Walter Benjamin, 148–84. Frankfurt am Main: Suhrkamp.

Benjamin, Walter. (1939) 1969. "Geschichtsphilosophische Thesen." In *Illuminationen: Ausgewählte Schriften*, edited by Walter Benjamin, 268–79. Frankfurt am Main: Suhrkamp.

Bijker, Wiebe E., Thomas P. Hughes, and Trevor J. Pinch, eds. (1987) 2001. *The Social Construction of Technological Systems: New Directions in the Sociology and History of Technology*. Cambridge, MA: MIT Press.

Björkman, Lisa. 2015. *Pipe Politics, Contested Waters: Embedded Infrastructures of Millennial Mumbai*. Durham, NC: Duke University Press.

Bloor, David. (1976) 1991. *Knowledge and Social Imagery*. London: University of Chicago Press.

Blumenberg, Hans. 1963. "Lebenswelt und Technisierung unter Aspekten der Phänomenologie." *Filosofia* 14, no. 4: 855–84.

Blumenberg, Hans. (1966) 1983. *The Legitimacy of the Modern Age*. Cambridge, MA: MIT Press.

Blumenberg, Hans. 1975. *Die Genesis der kopernikanischen Welt*. Frankfurt am Main: Suhrkamp.

Blumenberg, Hans. 1986. *Lebenszeit und Weltzeit*. Frankfurt am Main: Suhrkamp.

Blumenberg, Hans. 2010. *Theorie der Lebenswelt*. Berlin: Suhrkamp.

Blumenberg, Hans. (1963) 2020. "Phenomenological Aspects on Life-World and Technization." In *History, Metaphors, Fables: A Hans Blumenberg Reader*, edited by Hannes Bajohr, Florian Fuchs, and Joe Paul Kroll, 358–99. Ithaca, NY: Cornell University Press.

Boltanski, Luc, and Laurent Thévenot. (1991) 2006. *On Justification: Economies of Worth*. Princeton, NJ, and Oxford: Princeton University Press.

Boyer, Dominic. 2005. *Spirit and System: Media, Intellectuals, and the Dialectic in Modern German Culture*. Chicago: University of Chicago Press.

Breckenridge, Keith. 2016. "Biometric Capitalism." Technosphere Magazine, November 15. https://technosphere-magazine.hkw.de/p/Biometric-Capitalism-tAQgbSspe qhckkakBq8b5h.

Bueger, Christian, and Frank Gadinger. 2018. *International Practice Theory*. Cham: Springer International Publishing with Palgrave Macmillan.

Capek, Karel. (1920) 2024. *R.U.R. and the Vision of Artificial Life*, edited by Jitka Cejkova. Cambridge, MA: MIT Press.

Cassirer, Ernst. (1930) 2012. "Form and Technology." In *Ernst Cassirer on Form and Technology: Contemporary Readings*, edited by Aud Sissel Hoel and Ingvild Folkvord, 15–52. New York: Palgrave Macmillan.

Cooper, Frederick. 2002. *Africa since 1940: The Past of the Present*. Cambridge, UK: Cambridge University Press.

Crapanzano, Vincent. 2004. *Imaginative Horizons: An Essay in Literary-Philosophical Anthropology*. Chicago: University of Chicago Press.

Czarniawska, Barbara, and Bernward Joerges. 2020. *Robotisation of Work?* Cheltenham, Glos, and Northampton, MA: Edward Elgar Publishing.

Czarniawska-Joerges, Barbara. 1992. *Exploring Complex Organizations: A Cultural Perspective*. London: Sage.

Davidson, Donald. 1974. "On the Very Idea of a Conceptual Scheme." In *Inquiries into Truth and Interpretation*, edited by Donald Davidson, 183–98. Oxford: Clarendon Press.

De Laet, Marianne, and Annemarie Mol. 2000. "The Zimbabwe Bush Pump: Mechanics of a Fluid Technology." *Social Studies of Science* 30, no. 2: 225–63.

Deleuze, Gilles, and Félix Guattari. (1980) 1987. *A Thousand Plateaus: Capitalism and Schizophrenia*. Minneapolis: University of Minnesota Press.

DFG (Deutsche Forschungsgemeinschaft). 2019. "SPP 1448: Adaptation and Creativity in Africa. Technologies and Significations in the Production of Order and Disorder." https://gepris.dfg.de/gepris/projekt/128797734?language=en.

Diagne, Souleymane Bachir. 1989. *Boole: 1815–1864: l'oiseau de nuit en plein jour. Un savant, une époque.* Paris: Belin.

Diagne, Souleymane Bachir. 2013. «On the Postcolonial and the Universal.» *Rue Descartes* 78, no. 2: 7–18.

DiMaggio, Paul J., and Walter W. Powell. 1983. "The Iron Cage Revisited: Institutional Isomorphism and Collective Rationality in Organisational Fields." *American Sociological Review* 48, no. 2: 147–60.

Fanon, Frantz. (1961) 2004. *The Wretched of the Earth*, with introductions by Jean-Paul Sartre and Homi K. Bhabha. New York: Grove Press.

Feenberg, Andrew. 1996. "Marcuse or Habermas: Two Critiques of Technology." *Inquiry: An Interdisciplinary Journal of Philosophy* 36, no. 1: 45–70.

Feenberg, Andrew. 1999. *Questioning Technology*. London: Routledge.

Furlong, Kathryn. 2014. "STS beyond the 'Modern Infrastructure Ideal': Extending Theory by Engaging with Infrastructure Challenges in the South." *Technology in Society* 38: 139–47.

Gandy, Matthew. 2006. "Planning, Anti-Planning, and the Infrastructure Crisis Facing Metropolitan Lagos." In *Cities in Contemporary Africa*, edited by Martin J. Murray and Garth A. Myers, 247–64. New York: Palgrave Macmillan.

Gandy, Matthew. 2014. *The Fabric of Space: Water, Modernity, and the Urban Imagination*. Cambridge, MA: MIT Press.

Habermas, Jürgen. 1968. *Technik und Wissenschaft als "Ideologie"*. Frankfurt am Main: Suhrkamp.

Habermas, Jürgen. (1981) 1989. *The Theory of Communicative Action. Volume 2: Lifeworld and System: A Critique of Functionalist Reason*. Boston: Polity Press.

Halbwachs, Maurice. (1939) 1991. *Das kollektive Gedächtnis: Fischer, Frankfurt am Main 1991*. Frankfurt am Main: Fischer.

Harding, Sandra. 1992. "Rethinking Standpoint Epistemology: What is 'strong objectivity'?" *The Centennial Review* 36 (3): 437–70.

Heidegger, Martin. (1927) 2009. *Being and Time*. Malden, MA: Blackwell.

Heidegger, Martin. (1962) 2010. "The Question Concerning Technology." In *Technology and Values*, edited by Craig Hanks, 99–113. Malden, MA: Blackwell.

Hennion, Antoine, and Bruno Latour. 1996. "L'art, l'aura et la distance selon Benjamin, ou comment devenir célèbre en faisant tant d'erreurs à la fois ..." *Les cahiers de médiologie* 1.

Horkheimer, Max. 1947. *Eclipse of Reason*. New York: Oxford University Press.

Hountondji, Paulin. (1983) 1990. "Scientific Dependence in Africa Today." *Research in African Literatures* 21, no. 3: 5–15.

Hubig, Christoph. 2013. "Technik und Lebenswelt." In *Technik: Zeitschrift für Kulturphilosophie*, edited by Ralf Konersmann and Dirk Westerkamp, 255–69. Hamburg: Meiner.

Husserl, Edmund. (1934–37/1954) 1970. *The Crisis of European Sciences and Transcendental Phenomenology. An Introduction to Phenomenological Philosophy*. Evanston, IL: Northwestern University Press.

Ihde, Don. 1979. *Technics and Praxis*. Dordrecht and Boston: D. Reidel Pub. Co.

Ihde, Don. 1990. *Technology and the Lifeworld: From Garden to Earth*. Bloomington: Indiana University Press.

Ihde, Don. 2016. *Husserl's Missing Technologies*. New York: Fordham University Press.

Jackson, Michael. 2013. *Lifeworlds: Essays in Existential Anthropology*. Chicago: University of Chicago Press.

Jasanoff, Sheila, and Sang-Hyun Kim. 2015. *Dreamscapes of Modernity: Sociotechnical Imaginaries and the Fabrication of Power*. Chicago: University of Chicago Press.

Joerges, Bernward. 1988. *Technik im Alltag*. Frankfurt am Main: Suhrkamp.

Joerges, Bernward. 1996. *Technik – Körper der Gesellschaft: Arbeiten zur Techniksoziologie*. Frankfurt am Main: Suhrkamp.

Joerges, Bernward, and Barbara Czarniawska. 1998. "The Question of Technology, or How Organizations Inscribe the World." *Organization Studies* 19, no. 3: 363–85.

Jünger, Ernst. (1957) 1961. *The Glass Bees*. New York: Noonday Press.

Kaerlein, Timo. 2013. "Playing with Personal Media: On an Epistemology of Ignorance." *Culture Unbound: Journal of Current Cultural Research* 5, no. 4: 651–70.

Kaufmann, Matthias, and Richard Rottenburg. 2012. "Translation als Grundoperation bei der Wanderung von Ideen." In *Kultureller und sprachlicher Wandel von Wertbegriffen in Europa: Interdisziplinäre Perspektiven. Akten der internationalen Abschlusstagung zum Projekt "Normen- und Wertbegriffe in der Verständigung zwischen Ost- und Westeuropa"*, edited by Rosemarie Lühr, Natalia Mull, Jörg Oberthür, and Hartmut Rosa, 219–32. Bern: Peter Lang.

Kline, Ronald, and Trevor Pinch. 1996. "Users as Agents of Technological Change: The Social Construction of the Automobile in Rural America." *Technology and Culture* 37: 763–95.

Konersmann, Ralf, and Dirk Westerkamp, eds. 2013. *Zeitschrift für Kulturphilosophie 2013/2: Technik*. Hamburg: Meiner.

Koselleck, Reinhart. (1979) 2004. *Futures Past: On the Semantics of Historical Time*. New York: Columbia University Press.

Kracauer, Siegfried. 1960. *Theory of Film: The Redemption of Physical Reality*. New York: Oxford University Press.

Larkin, Brian. 2013. "The Politics and Poetics of Infrastructure." *Annual Review of Anthropology* 42, no. 1: 327–43.

Latour, Bruno. 1991. "Technology Is Society Made Durable." In *A Sociology of Monsters: Essays on Power, Technology and Domination*, edited by John Law, 103–31. London: Routledge.

Latour, Bruno. 1992. "Where are the Missing Masses? The Sociology of a Few Mundane Artifacts." In *Shaping Technology/Building Society: Studies in Sociotechnical Change*, edited by Wiebe E. Bijker and John Law, 225–58. Cambridge, MA: MIT Press.

Latour, Bruno. 1999. *Pandora's Hope: Essays on the Reality of Science Studies*. Cambridge, MA: Harvard University Press.

Law, John. 1994. *Organising Modernity*. Oxford: Blackwell.

Law, John. 2004. *After Method: Mess in Social Science Research*. London and New York: Routledge.

Law, John. 2008. "Actor-Network Theory and Material Semiotics." In *The New Blackwell Companion to Social Theory*, edited by Bryan S. Turner, 141–58. Oxford: Blackwell.

Lem, Stanisław. (1962) 1970. *Solaris*. New York: Walker.

Livingston, Julie. 2019. *Self-Devouring Growth: A Planetary Parable as Told from Southern Africa*. Durham, NC: Duke University Press.

Luhmann, Niklas. (1990) 1994. Die Wissenschaft der Gesellschaft. Frankfurt am Main: Suhrkamp.

Lynch, Michael. 2013. "Ontography: Investigating the Production of Things, Deflating Ontology." *Social Studies of Science* 43, no 3: 444–62.

Lyons, Kristina, Juno Salazar Parreñas, and Noah Tamarkin. 2017. "Engagements with Decolonization and Decoloniality in and at the Interfaces of STS." *Catalyst: Feminism, Theory, Technoscience* 3, no. 1: 1–47. http://www.catalystjournal.org.

Marcuse, Herbert. 1964. *The One-Dimensional Man: Studies in the Ideology of Advanced Industrial Society*. Boston, MA: Beacon Press.

Marx, Karl. (1858) 1983. *Grundrisse*, edited by Karl Marx and Friedrich Engels. Vol. 42: *Marx-Engels-Werke (MEW)*. Berlin: Dietz Verlag.

Mauss, Marcel. (1934) 1950. "Les techniques du corps." In *Sociologie et anthropologie*, 365–86. Paris: Presses Universitaires du France.

Mavhunga, Clapperton Chakanetsa. 2017. *What Do Science, Technology, and Innovation Mean from Africa?* Cambridge, MA: MIT Press.

Mavhunga, Clapperton Chakanetsa. 2018. *The Mobile Workshop: The Tsetse Fly and African Knowledge Production*. Cambridge, MA: MIT Press.

Mbembe, Achille. 1992. "The Banality of Power and the Aesthetics of Vulgarity in the Postcolony." *Public Culture* 4, no. 2: 1–30.

Mbembe, Joseph-Achille. 2016. *Politiques de l'inimitié*. Paris: la Découverte.

Merton, Robert King. 1993. *On the Shoulders of Giants: A Shandean Postscript*. Chicago: University of Chicago Press.

Miescher, Stephan. 2022. *A Dam for Africa: Akosombo Stories from Ghana*. Bloomington: Indiana University Press.

Moran, Dermot. 2000. *Introduction to Phenomenology*. New York: Routledge.

Mudimbe, Valentin-Yves. (1982) 2023. *The Scent of the Father: Essay on the Limits of Life and Science in sub-Saharan Africa*. Cambridge and Hoboken, NJ: Polity.

Mumford, Lewis. (1934) 1963. *Technics and Civilisation*. New York: Harcourt.

Orwell, George. 1945. *Animal Farm: A Fairy Story*. London and Edinburgh: Secker and Warburg.

Osseo-Asare, Abena Dove Agyepoma. 2019. *Atomic Junction: Nuclear Power in Africa after Independence*. Cambridge, UK and New York, NY, USA: Cambridge University Press.

Osterhammel, Jürgen. 1995. *Kolonialismus: Geschichte, Formen, Folgen*. München: Beck.

Oudshoorn, Nelly, and Trevor Pinch. 2003. *How Users Matter: The Co-construction of Users and Technologies*. Cambridge, MA: MIT Press.

Pickering, Andrew. 1995. *The Mangle of Practice: Time, Agency, and Science*. Chicago: University of Chicago.

Pinch, Trevor, and Wiebe E. Bijker. 1987. "The Social Construction of Facts and Artefacts: Or How the Sociology of Science and the Sociology of Technology Might Benefit Each Other."" In *The Social Construction of Technological Systems*, edited by Wiebe E. Bijker, Thomas P. Hughes, and Trevor J. Pinch, 17–50. Cambridge, MA: MIT Press.

Pollock, Neil, Robin Williams, and Luciana D'Adderio. 2016. "Global Software and its Provenance." *Social Studies of Science* 37 (2): 254–280.

Rammert, Werner. 1999. "Relations That Constitute Technology and Media That Make a Difference: Toward a Social Pragmatic Theory of Technicisation." *Techné: Research in Philosophy and Technology* 4, no. 3: 165–77.

Riedke, Eva. 2023. "A Solar Off-Grid Software: The Making of Infrastructures, Markets and Consumers 'Beyond Energy'." In *Digitisation and Low-Carbon Energy Transition*, edited by Siddharth Sareen and Katja Müller, 31–53. Basingstoke, Hants: Palgrave Macmillan.

Riedke, Eva, and Catherine Adelmann. 2022. "The Good Payers: Exploring Notions of Ownership in the Sale of Pay-as-You-Go Solar Home Systems." *Energy Research and Social Science*, 92: 102773.

Rodney, Walter. 1972. *How Europe Underdeveloped Africa*. London and Dar-es-Salaam: Bogle-L'Ouverture Publications.

Rogers, Everett M. (1962) 1995. *Diffusion of Innovations*. New York and London: The Free Press.

Rorty, Richard. 1989. *Contingency, Irony, and Solidarity*. Cambridge, UK and New York, NY, USA: Cambridge University Press.

Rosenberg, Nathan, and Manuel Trajtenberg. 2001. "A General Purpose Technology at Work: The Corliss steam engine in the late 19th Century US." NBER Working Paper Series. National Bureau of Economic Research, Cambridge, MA.

Rottenburg, Richard. 1994. "'We Have to Do Business as Business Is Done!' Zur Aneignung formaler Organisation in einem westafrikanischen Unternehmen." *Historische Anthropologie* 2, no. 2: 265–86.

Rottenburg, Richard. 1996. "When Organization Travels: On Intercultural Translation." In *Translating Organizational Change*, edited by Barbara Czarniawska and Guje Sevón, 191–240. Berlin and New York: de Gruyter.

Rottenburg, Richard. 2003. "Crossing Gaps of Indeterminacy: Some Theoretical Remarks." In *Translation and Ethnography: The Anthropological Challenge of Intercultural Understanding*, edited by Tullio Maranhao and Bernhard Streck, 30–43. Tucson: The University of Arizona Press.

Rottenburg, Richard. (2002) 2009. *Far-Fetched Facts: A Parable of Development Aid*. Cambridge, MA: MIT Press.

Rottenburg, Richard. 2014. "Experimental Engagements and Metacodes." *Common Knowledge* 20, no. 3: 540–48.

Said, Edward William. (1978) 2003. *Orientalism*. New York: Vintage Books.

Schütz, Alfred. (1932) 1967. *The Phenomenology of the Social World*. Evanston, IL: Northwestern University Press.

Schütz, Alfred, and Thomas Luckmann. 1974. "The Life-World as the Unexamined Ground of the Natural World View." In *Structures of the Life-World*, edited by Alfred Schütz and Thomas Luckmann, 3–8. Evanston, IL: Northwestern University Press.

Senghor, Leopold Sédar. 1956. "The Spirit of Civilisation or the Laws of African Negro Culture." *Presence Africaine*, no. 8–10: 51–64.

Serres, Michel, ed. (1989) 1995. *A History of Scientific Thought: Elements of a History of Science*. Oxford and Cambridge, MA: Blackwell.

Simone, AbdouMaliq. 2004. "People as Infrastructure: Intersecting Fragments in Johannesburg." *Public Culture* 16, no. 3: 407–29.

Sloterdijk, Peter. 2023. *Die Reue des Prometheus: Von der Gabe des Feuers zur globalen Brandstiftung*. Berlin: Suhrkamp.

Star, Susan Leigh. 1999. "The Ethnography of Infrastructure." *American Behavioral Scientist* 43, no. 3: 377–91.

Stengers, Isabelle. (2009) 2015. *In Catastrophic Times: Resisting the Coming Barbarism*. Lüneburg: Open Humanities Press/Meson Press.

Strauss, Anselm. 1978. "A Social World Perspective." *Studies in Symbolic Interaction* 1: 119–28.

Thiong'o, Ngũgĩ wa. (1981) 2005. *Decolonising the Mind: The Politics of Language in African Literature*. Oxford and Nairobi: James Currey and EAEP.

Trovalla, Eric, and Ulrika Trovalla. 2015. "Infrastructure as a Divination Tool: Whispers from the Grids in a Nigerian City." *City* 19, no. 2–3: 332–43.

Tuck, Eve, and K. Wayne Yang. 2012. "Decolonization Is Not a Metaphor." *Decolonization: Indigineity, Education and Society* 1, no. 1: 1–40.

UNESCO (UN Educational, Scientific and Cultural Organization). 1974. *Science and Technology in African Development*. Paris: The UNESCO Press.

Van der Waerden, Bartel L. 1985. *A History of Algebra: From al-Khwarizmi to Emmy Noether*. New York: Springer.

Van Laak, Dirk. 2004. *Imperiale Infrastruktur: Deutsche Planungen für eine Erschließung Afrikas 1880–1960*. Paderborn: Schöningh.

Von Albertini, Rudolf, and Albert Wirz. (1976) 1982. *European Colonial Rule, 1880–1940: The Impact of the West on India, Southeast Asia, and Africa. Vol. 10: Contributions in Comparative Colonial Studies*. Westport, CT: Greenwood Press.

Von Schnitzler, Antina. 2016. *Democracy's Infrastructure: Techno-Politics and Protest after Apartheid*. Princeton, NJ: Princeton University Press.

Whyte, William Foote. 1948. *Human Relations in the Restaurant Industry*. New York: McGraw-Hill.

Wynne, Brian. 1988. "Unruly Technology: Practical Rules, Impractical Discourses and Public Understanding." *Social Studies of Science* 18, no. 1: 147–67.

CHAPTER 2

PAYGo Water Dispensers and the Lifeworlds of Marketisation

Christiane Tristl and Marc Boeckler

1 Introduction: Making Sense of Water Provision

In Kondo, a village in rural Kenya, a woman arrives at a water kiosk which is a common source of water supply for the poorer population in the Global South. She places her yellow twenty-litre jerry can underneath the water tap. On a ribbon around her neck, she carries a bright blue water card featuring a graphic symbol of a white drop of water. She places the card on the equally bright blue slot in the water dispenser, which has been installed in the window frame of the kiosk, presses the now flickering water button and administers the sudden flow of water into the reutilised cooking oil container. Only a few weeks ago, this mechanism was manually operated by a kiosk attendant selling water from behind the kiosk window. As her jerry can fills with water, the woman tracks the credit balance on her water card by monitoring the digital interface on the dispenser. As soon as the jerry can is full, she removes the card, and the water flow stops. The woman picks up the jerry can and leaves for her next destination.

PAYGo water dispensers combine prepaid services (pay-as-you-go) with mobile money applications, and remote monitoring systems. They are being installed at water points or in water kiosks in the Global South, and in so doing, replacing human kiosk attendants. In contrast to the large-scale infrastructure projects of earlier development eras and their associated ambitions to modernise national water supply, these small technical devices directly serve individuals or communities in infrastructurally marginalised parts of the world. PAYGo dispensers have been promoted by various bodies of international organisations, including philanthrocapitalist foundations, as a market-based solution for failing communal water systems. Their distribution is driven by the moral obligation to provide sustainable water access founded on philanthrocapitalist principles that assume the realisation of such sustainability through a market orientation.

As an infrastructural market device, the water dispenser is designed to slip seamlessly into the lifeworld of its users. As our introductory vignette suggests, the machine does exactly that: at the touch of a button, a can of water is filled, allowing the complicated metrological process of measuring and paying to fade into the background. However, even where the automated water dispenser functions properly and it travels easily from one location to another, has it really become an element of the lifeworld of Kondo?

It is here that we want to turn to the phenomenological writings of Hans Blumenberg to develop a deeper understanding of the process of technicisation. According to Blumenberg, technicisation is *Sinnverzicht*, that is, a "relief from the burden to make sense" or a "relief from the burden to grasp the full meaning," or a "waiver of meaning"(see Rottenburg et al., introduction to this volume). We seek to bring Blumenberg's ([1963] 2010) reworking of the phenomenological concept of the lifeworld into conversation with Science and Technology Studies (STS), particularly through the work of Michel Callon (1998c). Focusing on moments of "breakdown" (Akrich 1992) when PAYGo water dispensers travel from one place to another, we are particularly interested in empirical situations in which the "waiver of meaning" collapses. We use our empirical material to demonstrate that the everyday infrastructuring of life via PAYGo dispensers is not about water provision but rather about the marketisation of society.

In what follows, we will first (sections 1 and 2) situate the dispenser in the political economy of water provision by outlining a brief genealogy of water policies and related technologies in development cooperation. Second (section 3), we will introduce our view on markets as sociotechnical *agencements* and Blumenberg's notion of *Sinnverzicht*. Developing *de-scription* as an empirical approach to moments of disruption, interference, and confusion, will allow us to question the portability of the dispenser's meaning. Finally (section 4), we will present our empirical material on the displacement of a water dispenser from the European engineering company, where it was created, to East African villages. We will do so by analysing a project that has been funded by a philanthrocapitalist foundation from the United Kingdom and is managed by a major international NGO active in the Water, Sanitation and Hygiene (WASH) sector. We will draw on interviews with the manufacturers and designers of the PAYGo model and on ethnographic material tracing its implementation in Kenya from 2015 to 2018.

2   The Advent and Reoccurrence of Market Principles in the Water Sector

A global shift towards the treatment of water as an economic good and towards the principle of cost recovery in water systems took place in 1992 with the declaration of the Dublin Principles. As a counterpart to the privatisation of urban water utilities, local communities in rural areas were now expected to independently manage their water systems (Black 1998). In the early development era, starting after the Second World War, water supply was generally seen as a state responsibility, to be provided free of charge. Large-scale infrastructure projects of this era—paradigmatically embodied in the water dam—were characterised by poor adaption of highly sophisticated technological infrastructure into local contexts. As a result, these infrastructure projects were met with increasing criticism by grassroots organisations, which eventually led to a new paradigm in the 1970s that foregrounded alternative, "appropriate" technologies that were more adaptable to the respective local contexts (Cherlet 2014).

In the water sector, the handpump spearheaded the adapted technology phase. As a key exemplar of an appropriate water technology, Marianne de Laet and Annemarie Mol (2000) describe the Zimbabwe Bush Pump as a modest and fluid device that is adaptable to local behaviours and environments, fostering community cohesion rather than individualisation. While handpumps were designed to be operated and maintained by local communities, governments had a great responsibility in standardising pumps to ensure local production in order to guarantee a nationwide availability of spare parts. In being "democratic," adapted technologies were meant to "empower" communities with control over their own water-related affairs (Black 1998, 20). The paradigm of community water-management, together with the use of appropriate technologies, was reinterpreted in the context of neoliberal policies during the period of structural adjustment in the 1980s and 1990s. While the state was still assigned a significant role in larger maintenance works and community training, following the Dublin Principles, community systems were expected to shift to full-cost-pricing-based tariffs, with the aim of full cost recovery (IRC 1992). Community involvement was now expressed in financial terms in the sense of a "willingness to pay" for a valued service (Black 1998). This "demand responsive" approach also marked a notable shift in responsibilities. After "rolling back" African states, civil society, which was expected to flourish with the support of skilled NGOs, was put forward as the main actor to see through the implementation of community projects (Ferguson 2007). Governments served only as facilitators of community-managed water systems (Black 1998).

However, in 2015, one in four handpumps in sub-Saharan Africa (a total of 175 000) were non-functional. It has therefore been argued that community-managed water systems were never actually realised as intended (Foster et al. 2019). Moreover, with the influx of professionalised NGOs during the 1980s and 1990s, the not-for-profit sector incrementally took over state functions (Ferguson and Gupta 2002) and NGOs emerged as de facto water service providers. Water systems continued to collapse soon after installation and were in constant need of non-governmental repair (Harvey and Reed 2003).

Under the output orientation of the Millennium Development Goals (MDG) and its directives for "access" to water supply (Aqua for All et al. 2017, 5) as an immediate remedy to the disastrous effects of structural adjustment programmes (Roy 2010), this process accelerated. Yet, there were a number of other reasons for the failure of water systems. The concept of community management, for example, was often applied in a top-down manner that romanticised communities (Bakker 2008); policies of trade liberalisation interfered with the national standardisation of handpumps; boreholes were installed incorrectly by NGOS due to a lack of hydrological knowledge (Harvey and Reed 2003). The latest policy shift in the field of water supply circumvents this complex fabric of larger structural and political questions. Instead, by proposing PAYGo systems as a solution, the failure of water systems now falls on the individual.

3   PAYGo Dispensers, Philanthrocapitalism and the New
       Technopolitics of Development

The support for PAYGo water dispensers ranges from traditional donors, such as the International Water and Sanitation Centre (IRC), an influential water, sanitation, and hygiene think-and-do-tank (Smits et al. 2016), or the World Bank (Ikeda and Liffiton 2019), to financial inclusion think tanks, the telecommunications industry (Waldron et al. 2019), and philanthropic foundations (Aqua for All et al. 2017). The narratives of these organisations reduce the manifold reasons for why community-managed water systems break down to "inadequate operation and maintenance" (Smits et al. 2016, 1), which they attribute to a lack of resources due to so-called non-revenue water—the provision of water free of charge. The result would be a negative cycle of "construction-use-breakdown-construction" (Smits et al. 2016): when water systems break down soon after installation and communities have no budget for repairs, NGOS, as implementing agencies, start over again.

Moreover, these organisations argue that past approaches have neglected maintenance, repair, and the overall sustainability of water systems and focused too heavily on the provision of mere "access" to water supply (Aqua for All et al. 2017, 5). Therefore, the Sustainable Development Goals (SDG) demand an increasing emphasis on the "sustainability", that is, *financial* sustainability—of water systems (Smits et al. 2016). The orientation of the SDGS converges with philanthrocapitalism's market-oriented logic. For philanthrocapitalists—for which the *Bill and Melinda Gates Foundation* is the best-known example—any kind of charitable engagement, from almsgiving to corporate philanthropy, is an unsustainable one-time donation (McGoey 2016). While regarding private sector engagement as superior to conventional development aid, their quest is "transformative" in the sense that only "strategic" and "effective" giving (Bishop 2006), what Brooks et al. (2009, 10) refer to as "connecting to the market," is sustainable and will ultimately benefit "the poor."

For these humanitarian and development actors, PAYGo dispensers are best suited to improve water supply for "the poor" in two ways. First, PAYGo devices are supposed to eliminate "human error" or "fraud" (GSMA 2018, 8). As such, they reduce the "commercial losses" (Ikeda and Liffiton 2019, 16) of water systems and create an "equitable system in which status, relationships or income [do] not affect access to water" (GSMA 2018, 8). Together with the remote monitoring systems, the devices should cater for efficient revenue collection and full transparency of all transactions made, thereby improving the management of water systems. Using them to enforce economic rationality, their proponents argue that they will ensure the profitability and upkeep of water systems. Second, having been tested in the field, the user-friendly and intuitive interfaces of the devices should also guarantee easy access for the targeted population (Waldron et al. 2019). In contrast to monthly billing, their application is argued to be financially inclusive, "affordable[,] and convenient for those with irregular incomes" (GSMA, 2017, 5), since users can conveniently recharge the water cards or tokens supplied with the devices at any time using their mobile money application.

PAYGo water dispensers can be associated with a broader trend of "little development devices" (Collier et al. 2017), which directly address the needs of the "infrastructurally marginal" (Collier et al. 2017). First, small development devices depart from the grand infrastructure projects of the post-war period which were primarily driven by the state as the major force behind structural change and modernisation of nations. Instead, small development devices operate on a micro-scale, often addressing the level of the individual or community, and are often offered or driven by private sector companies and/or philanthropic foundations (Collier et al. 2017). When analysing "the new technopolitics of development," Adam Fejerskov (2017) has observed that the new philanthrocapitalists are pushing development policies that combine individualisation with the market rationale, and the idea of societal progress through innovative technologies. In his view, these approaches consider "poverty to be an engineering problem, in which the innovation of new technologies is the main road to solving human misery" (Fejerskov 2017, 954). Second, devices such as solar lights (Cross 2013), financial inclusion apps (Schwittay 2014), or water filters (Redfield 2015; Irani 2019), are often developed in the field with the support of anthropologists and adjusted to the environments in which they will be used.

What is remarkable regarding these new approaches in the water sector is the persistent belief in market orientation as universal and the only solution to development problems. For development cooperation in general, Marc Boeckler and Christian Berndt (2013) observed that rather than weakening the neoclassical project, past market failures serve as a background to justify new, intensified, and more intrusive market-oriented approaches. In contradistinction to the belief that the hidden hand of the free market works best without support, the designers of PAYGo water dispensers realise the importance of infrastructuring the marketisation of water in order to engineer individual behaviour to fit the conditions of market rationality (Berndt and Boeckler 2016).

Few studies to date have looked at water devices in the context of market-based development. As a classic Millennium Development Goal (MDG) technology, the lifestraw—described by Peter Redfield as a water filtering device to satisfy immediate needs beyond infrastructure—also fosters radical individualisation. Siding with the humanitarian end of small devices for "the poor," it embodies an "alternative model of corporate liberal care" (Redfield 2015, 16) as it is financed through carbon markets and distributed free of charge. Antina von Schnitzler (2008) has shown in her analysis of infrastructural politics in Johannesburg that new modes of calculability and budgeting were enacted with the deployment of prepaid metres. As the chapter elucidates, PAYGo dispensers generate "budgeting individuals" in a similar fashion. They can be regarded as rigid machines of marketisation that will correct behavioural deficiencies.

4   Marketisation, De-scription and the "Waiver of Sense"

Divergent from orthodox or neoclassical economics that claims that economics is an abstraction of the economy, social studies of marketisation argue that economics is performative and brings about the very reality that it purports to describe (Callon 2007). These studies assume that any entity is always embedded in a heterogeneous network of materials and relationships. Buyers or sellers are caught up in personal relationships, making rational calculations in the sense of neoclassical economics impossible. The demarcation of an economic good relies on various measuring devices. Some "framing" (Callon 1998a) is necessary for any commercial transaction to take place. Such framing disentangles the relationships that will be taken into account from the relationships that will be ignored. Calculative agency, then, is the outcome of a configuration process that involves not only economic models but also calculating tools and is therefore distributed within such sociotechnical configurations (Callon 1998b).

The dispenser links two processes. First, PAYGo dispensers render indefinite masses of water calculable and commensurable. Water is translated into measured litres and a set price is attached to this unit of measurement. Second, the severing of social linkages and containment of potential overflows builds upon a process of translation that displaces the conditions of economic modelling into society (Latour 1983; MacKenzie, Muniesa, and Siu 2007). By equipping each water user with a smart card, individual budgets are to be managed and individualised transactions to be performed at the dispenser interface. Eventually, the water dispenser is a "calculative articulation of things" (Muniesa 2014, 39), with the creation of rational budgeting and water calculating consumers at its centre. Market-oriented water policies are materialised within the machine that imagines society as a set of atomised individuals. The dispenser represents a figuration of "radical neo-individualism" (Berndt and Boeckler 2016, 22), or "roll-in-neoliberalism" (Berndt and Boeckler 2017), that shifts its target of policy intervention from the market to the individual by attributing poverty not to structural inequality but to erroneous behaviour.

Borrowing from Gilles Deleuze, Callon has conceptualised markets as sociotechnical *agencements* explained as materially heterogeneous arrangements that organise the circulation of goods together with the property rights attached to them, through the contradictory encounter of quantitative and qualitative valuations. Agencements are self-evidentiary. For Callon (2007, 320), there is nothing outside of the agencement: "There is no need for further explanation, because the construction of its meaning is part of an agencement." It is here that we turn to Blumenberg ([1963] 2010, 198) and his revised concept of phenomenology's lifeworld as the *Universum vorgegebener Selbstverständlichkeiten*, the "universe of the given, of what is always unquestionably available," which he relates to the process of *Technisierung*.

While studies in ANT have shown that certain tasks can be inscribed into or delegated to technologies (Latour 1992; Akrich 1992), Blumenberg ([1963] 2010) describes the delegation to technology, or "technicisation," as the transformation of originally active sense-making into something that can be passed on without carrying along its raison d'être, its foundational meaning. "The loss of meaning" that Edmund Husserl bemoaned is in fact a purposeful or "self-imposed waiver of sense," *ein selbst auferlegter Sinnverzicht* (Blumenberg [1963] 2010, 216), that enables a relief from having to comprehend and prove historically achieved knowledge anew each time. Since human lives are limited, Blumenberg states that formalisation is necessarily an essential method for any advancement in knowledge, a prerequisite for science and development, and a condition of modernity. In opposition to the anti-technological Frankfurt School, the "colonisation of lifeworld" is nothing more than the mundane constitution of the lifeworld through practices of routinisation and standardisation (Rottenburg et al., introduction to this volume). However, technicisation as an everyday practice of *Sinnverzicht* is always situated (Graham and McFarlane 2014) and this situatedness raises the important question of the portability of the lifeworld.

According to Madeleine Akrich (1992), technologies embody "scripts," specific setups that are a result of how designers envisioned the place and people where the technology is to be implemented. Such a script maps certain roles for environments and users. However, the environment of implementation can never be fully predetermined by the script. Users and environments may come forward to play their intended roles, or they may not. Thus, a technical object may alter the way heterogeneous elements, such as technical components, users, customers, or maintenance personel, relate to the object and to each other. However, these elements may also be able to reshape the object in question and the way it might be used. As such, the boundary of the object does not reside within the technical object itself but is a *"consequence,"* as Akrich (1992, 206) describes, of a process of mutual adjustment between the object in question and these heterogeneous elements.

Conversely, tracing objects as they travel allows us to find "circumstances, in which the inside and the outside of objects are not well matched" (Akrich 1992, 207). Such moments of "crisis" or "breakdown" allow for the "de-scription" of technical objects (Akrich and Latour 1992, 259). For Akrich, *de-scription* is an empirical term that captures the iterative relationship between designers and users, but it can become an analytical concept that we as researchers can use. We can do "the opposite movement of the in-scription by the engineer, inventor, manufacturer, or designer" and instead "translate from things back to words" (Akrich and Latour 1992, 260). Regarding its deployment in the Global South, "de-scription" shares the conflation of an empirical condition with a methodological perspective of "infrastructural inversion" (Harvey, Jensen, and Morita 2017). For Geoffrey Bowker, *infrastructural inversion* is a methodological term that foregrounds infrastructure that, when functioning, recedes into the background (Bowker 1994). However, when breakdown is the norm, inversion becomes the modus operandi of infrastructure and researchers alike.

We want to utilise this method by tracing PAYGo water dispensers as they travel from their point of design to Kenya, in order to unearth their foundational meaning through the *de-scription* of moments of disruption. At the interface of science and technology studies and phenomenology, we can divide the de-scription of the water dispenser into two sets of inquiry. First, does the machine, after its displacement and material reassembly, function as intended by its designers? Are the relations of technical properties stable yet mobile? Do they travel well without rupturing? Is causality successfully fixed in matter while at the same time remaining mobile? As an "immutable mobile" (Latour 1987), the dispenser makes "the economy portable" through the economisation, abstraction, and the valuation of water (Muniesa 2014, 40). However, if "technology is society made durable" (Latour 1991), what kind of society has been materially fixed in the water dispenser and what kind of society does it seek to replace?

Second, we therefore need to ask how mobile the lifeworld of a technology like our dispenser is. Blumenberg ([1963] 2010, 211; our translation) pointed out that "the always-ready-to-hand, the at the touch of a finger triggerable and retrievable does not justify its existence." Instead, the technology appears already legitimised by being put into operation as part of a lifeworld. Being made self-evident as an element of a lifeworld, it "silences all questions as to whether this is necessary, sensible, humane, somehow justifiable" (Blumenberg [1963] 2010, 211; our translation). Regarding the travels of a water dispenser from a European design studio to rural Kenya, we question: Has the "waiver of meaning" (*Sinnverzicht*) survived the translation into novel settings, i.e. does the dispenser seamlessly enter into use? Or is the mobility of the lifeworld a more complicated endeavour because it needs to be woven into a new lifeworld by new waivers of meaning?

Following the methodology of *de-scription* as an attentiveness towards moments of disruption, misunderstanding, or simply confusion, we are looking for situations in which the automatic water dispenser loses its self-evidence, the machine is being questioned anew, the waiver of meaning is repealed and the *Urstiftungssinn*—the "original meaning"—is revealed.

5 The De-scription of a Market Device

5.1 *Designing an Intuitive User Interface*

Blumenberg ([1963] 2010, 210) described the example of the doorbell to illustrate the technicised lifeworld, which we translated into English: "With the touch of a button, the withdrawal of insight [in the most literal sense of looking in!] is celebrated: command and effect, order and product, will and work are moved together to the shortest distance." As if the dispenser had been designed to performatively bring Blumenberg into being, it closely follows the principle of the doorbell by "sinking back into the lifeworld" through the creation of a universe of the self-evident (Blumenberg [1963] 2010, 210; our translation).

The language that Joost, the lead engineer of the PAYGo dispensers, chose to explain the design of the dispenser's interface (Figure 2.1) were "extremely user-friendly," "intuitive to use," and "engaging".[1] The designer applied what is termed "natural user interfaces" in media studies and interface design (Wigdor and Wixon 2011). The goal is to create a technology that is self-explanatory and can immediately be integrated into the user's world. Johan, the designer of the dispenser, tried to solve the problem in the following way:

> I said the goal is that a mother [sic] explains to her girl [sic] how she can get water out of it and then the girl [sic] has to be able to get water out of it … The people can just throw their cards at the thing and then the button is blinking and then you push the button … The people have a blue card, they go to the blue thing at the machine and then they see a drop of water … So, it is just like with my children: they have a piece that fits into a box where the contours are star shaped and red and they have a brick that is star shaped and red.[2]

The designer's main goal was to realise a product that anybody, even children, could operate with little or no explanation: put the water card in the right slot and press the water button to make the water flow. Therefore the colour and water drop symbol on the card is the same as the holder in which to place the card. At the same time, the water button constantly blinks, sending out subtle affordances that invite the user to press it.

The design of the dispenser can be associated with what Timo Kaerlein (2013) terms "infantilisation" which is the trivialisation of mobile media through colourful icons, touchscreens, or customised apps. Drawing on Blumenberg's concept of technicisation, Kaerlein (2013) argues that in order for such devices to be inclusive and open to every user, they have to be familiar and self-explanatory, black boxing their often complex inner workings. Market-based development seeks to adjust itself to the presumed conditions of its customers and portraying "the poor" as in need of its adjusted products: for example, accepting illiteracy instead of promoting literacy, or accepting gender relations rather than problematising them. Most of the people who were observed engaging with the dispensers intuitively knew where to place their card. Reminiscent of Blumenberg's example of the doorbell, they

---

1   Interview with Joost, July 2016.
2   Interview with Johan, July 2017; our translation.

FIGURE 2.1   PAYGo dispensers with interfaces and smart cards
             PHOTOGRAPHY BY CHRISTIANE TRISTL

naturally followed the affordance imbued in the blinking water button. The operation of the dispenser was quickly internalised. At a first glance, the journey of the waiver of sense seemed to be successful. For the full de-scription of the lifeworld, however, we need to take a closer look. To this end, we investigate the implementation of PAYGo dispensers in Kondo, a village in the East of Nairobi, which is a particularly interesting case for our analysis due to its specific geographical and historical context.

5.2   The Morality of the Market

Kondo is located in Makueni County of Kenya, which belongs to Ukambani, the land of the Kamba people. The Kamba are the largest ethnic group in Kenya, accounting for almost 11 percent of the country's population. Ukambani has historically been associated with multiple crises such as drought and famine. In pre-colonial times, the people of Ukambani lived in accordance with flexible patterns of settlement and mobility that followed a mix of private and common property rights. Arrangements of mutual reciprocity in terms of mobilising support from family, clan, church, and other networks of mutual support have always played an important role in the organisation of Kamba societies, for instance as a mechanism to limit vulnerability during times of drought (Rocheleau, Benjamin, and Diang'a 1995). Colonialism, however, forced the Kamba into permanent sedentary settlement structures. In the recent past, Makueni has been regularly affected by drought-related, large-scale failure of crops, resulting in the distribution of food aid by the Kenyan government. The majority of the population depends on subsistence farming with casual labour as the major source of income (NDMA 2014). Thus, income flows are of immediate, irregular, and short-term in nature. With 64 percent, the poverty level is far greater than the Kenyan average of 45 percent (World Bank 2013).

The dispensers were installed in all water kiosks in Kondo. The kiosks cater for the supply of potable ground water, as they are connected to boreholes via metal pipes. Furthermore, earth dams collect water during the rainy seasons in March and April (with a short second season in November and December), which can be used for certain domestic uses as well as for irrigation and livestock. During the dry season, the only permanent source of water next to the boreholes is the heavily polluted Athi River.

The water system in Kondo was managed by a local committee of community members. While vendors sold water, the task of the committee was to manage the budget and organise repairs if necessary. However, the system lacked revenue for maintenance most of the time. Therefore, both the NGO and the philanthrocapitalist foundation that carried out the project pinned high hopes on the dispenser as the solution to an independent system that would finally ensure cost recovery. The reason for the low revenue was considered to be the inefficiency of the system due to corruption, irrationality, or simply incompetence on the part of the actors involved. The dispenser was now to correct deviations from the desired model, as Frank, an NGO worker, explained:

> It could also be that they [water vendors] give some free water. Because if your mother-in-law comes, or your sister, or your father, or friend, who says "oh, I don't have money, give me some water and I will bring money tomorrow," maybe you are tempted to give. The dispenser does not know any in-law, or brother, or friend. It doesn't understand language. It just understands a card, a water card with credits.[3]

By embodying economic rationality, the PAYGo dispensers mobilise assumptions about the behavioural deficiencies of human beings. Humans tend to be caught up in relationships, emotions, and temptations. Instead, PAYGo dispensers will cater for a fair, equitable, and morally stable relationship between water distribution and revenue collection. It is the secret of market transactions that they sever all the social and material ties of homo apertus,

---

3   Interview with Frank, NGO worker, June 2016.

thus discounting all externalities that are not part of the price of the item purchased. A market transaction is successful when a good has been disentangled from its context, changes hands, and the exchange between buyers and sellers is considered complete—without the addition of social obligations and other human responsibilities. What we could read as "the morality of the market" does not know father nor sister nor friend—it will treat everybody equally.

In Kondo, competing visions of morality can be found. One of the water kiosks was not yet equipped with a dispenser because it was on private property. When asking Ezekiel, the kiosk attendant, if he would deny people water without payment, he replied with laughter:

> [Laughs] You know, we are human beings [laughs] ... Machines can do that. But human beings—we can't. Yeah. When someone comes and tells you: "I need water but I pay afterwards," you just have to. You can't deny them. It is not right. Yes. It is not right. So, you just have to give them [water], and then after, maybe in the evening they bring [the money]. Or maybe they come the following day. In the morning, they pay. Then they fetch another one. That is how we handle these cases.[4]

This version of morality speaks to an environment that is characterised by its short-term nature and immediacy, by spontaneous events, and emergencies— be they caused by drought, lack of income, or general poverty. It is reminiscent of earlier drought adaptation systems in Ukambani, which were characterized by a remarkable flexibility in moving between places, but also in terms of mutual support. To bring about market morality through technicisation, that is, the delegation of economic rationality to technology, excludes the possibility of situation-specific reasoning. PAYGo dispensers will materially turn water into an economic good and bring about the transactional, rational, standardised distribution of water.

5.3   *Pay-as-you-drink*
The new constellation, which prohibits access to water without a recharged water card, necessitates practices of calculating and budgeting. As one water user explained:

> On Saturday, which is market day in Kondo, I will work during the day and by the end of the day I will get some amount, which is enough to

---

4 Interview with Ezekiel, water dispenser attendant, February 2017.

push me for the cost of water through the week. And the money will be recharged in the card. So every Saturday I take two hundred shillings to recharge the card specifically for water costs.[5]

"Earmarking" one's water budget particularises money (Zelizer 1997) and links it to specific uses and circulations, which can be carefully supervised. Thus, the water card emerged as an important material device for self-discipline. The rational management of one's water is distributed between the user and the card.

The ability to budget does not, however, only depend on obtaining a card but also on obtaining money to budget with. Users with irregular and unstable incomes often used their cards until the balance was zero and only then started to "search for money" to recharge the card. For example, as Rose explained, the new payment constellation created an extra burden for people with a limited budget: "It is not common in my home that I recharged the card to make water ready and there is no food. So, I prefer to have money somewhere, which can buy food so that the other small amount now has been put in card to buy water."[6] When faced with small or irregular budgets, how much money should the card be recharged with in order to still have an amount left for food? In Kondo, people never had the opportunity to sign monthly billing agreements with water service providers. Rather, they always had to pay immediately upon receiving water at the water kiosk. Narratives that emphasize the convenience of PAYGo for people with small, irregular incomes ignore two important characteristics of their situation. First, PAYGo necessarily demands from its users to extend credit to the water system before payment is possible, whereas for most users payment in cash would still be the most obvious and convenient form of payment. . Second, coping mechanisms such as users being granted credit at the water kiosk in case of emergencies, are foreclosed. Relationships of market exchange have been fully established in the sense of "Pay-as-You-Drink" (Waldron, Hwang, and Yeboah 2018). As the water system became disentangled from social relations, people without any budget to allocate to water, had to search for backup solutions such as asking their neighbours to borrow water. Or they had to substitute kiosk water with alternative, polluted solutions: "Now, when there is no money, we go to the dams and rivers ..."[7]

---

5   Interview with Benson, February 2017.
6   Interview with Rose, water dispenser user, January 2017.
7   Interview with Rose, water dispenser user, January 2017.

5.4   *MPesa and Sticky Water*

PAYGo dispensers mobilise assumptions about the universality of the mobile money application MPesa as well. Before the water cards can be recharged with MPesa, users first have to top up their MPesa accounts by depositing cash at an MPesa agent. There were certainly people in Kondo who fit into the imaginary of being avid MPesa users. Especially more business-oriented people always had credit in their MPesa accounts. Others only used MPesa for specific situations rather than on a daily basis, such as receiving remittances from relatives in Nairobi. In addition, while there are plenty of MPesa shops in Kenya, beyond urban agglomerations these shops are usually located at what is referred to as the "market", small sub-centres that serve vast areas with basic goods. In Kondo Market, there are many bright green coloured shops which signals the presence of an MPesa agent. However, boreholes and water kiosks in Kondo are spread out several kilometres away from the market. "Now the card takes me to the market," as Shadrack, another user, explained.[8]

The telco industry celebrates the entanglement of water and mobile money as a promising opportunity for market expansion: "There is no commodity more essential than water and if mobile money is connected to its daily procurement that can be a powerful and sticky introduction to mobile bill payments" (Waldron et al. 2019, 14). These visions, however, do not take into account the spatiality of digital infrastructures and mobile money applications. In Kondo, going *to* the market can be equated with walking *to* the market. The entanglement of water and mobile money in fact complicates a common aid-narrative which tells us that it is usually women and girls who have to walk far to reach the next water point. PAYGo dispensers drag MPesa users into the very "tyranny of geographers in defining space" that Bruno Latour (1996, 371) wanted to abandon in favour of "connectivity." Even though the water kiosk might be next door, to achieve connectivity, the user must walk to the market, recharge the card, return to the kiosk and fill the jerry can. What is more, it also ignores the situated temporalities of a place like Kondo, where in cases of small spending budgets, recharging the card has to be frequently repeated. PAYGo dispensers become a geographical exercise in relating topological connectivity of payment systems with the topographical distance to markets.

5.5   *The Negotiability of Twenty Litres*

Marketisation relies on a single space of calculative commensurability. Water and money must be brought to the same surface. The relational value of the

---

8   Interview with Shadrack, water dispenser user, February 2017.

product is assessed and then translated into a number, a price: "singular and comparable, and consequently calculable" (Callon and Muniesa 2005, 1236). In the case of the dispenser, there have been complicated metrological disputes before a pricing formula could be fixed.

Water dispensers promise full transparency and efficient management of water systems. Every transaction—dispensed litres and payment of the equivalent price—can be traced by the remote monitoring system. In its fluidity, water is not automatically tied to money. To establish a pricing formula, water has to be measured and subdivided into units. In the case of the dispenser, litres are attached to a non-negotiable price. When water flows through the dispenser, the flow sensor will calculate its flow rate. The result is shown on the screen in the form of a rising number of litres and a decreasing credit balance on the water card. However, far from *being universal*, litres have *to be made universal* as the product of modern scientific culture.

Standing in front of the interface of the dispenser in Kondo and looking at its upper right corner, there is the *odd* number of 0.15. Users usually did not know what to make of this number. It is the price per litre. In order for the water data to be transferable across distance, and for the dispensers themselves to be transferable across contexts, it has to operate with some presumably universally understandable metrologies. How did this number come about? In Kondo, water is not measured in litres, but in twenty litre jerry cans (Figure 2.2). The price for a jerry can of water is KES3. With the introduction of the dispenser, it was decided that the price for a jerry can of water should remain unchanged. Therefore, the price per litre was set at KES0.15. However, people fill up their jerry cans at the water kiosk and then have to walk long distances to carry them home. To maximise efficiency, people make the canisters as full as possible. For this purpose, the jerry can is tilted backwards so that the opening is at maximum height. As such, a twenty litre jerry can fit twenty-two, or even twenty-three litres.

"At first, we could not measure accurately. But now it is being measured accurately. Now the dispenser is telling the truth", as Josphat, the chairman of one of the water committees, explained.[9] The idea of the final implementation of accurate measurement led to a discussion about the "normal" level of water amongst users:

Jack: Three shillings is at the normal level.

Christiane: Who says that this is the normal level?

---

9   Interview with Josphat, water committee chairman, March 2017.

FIGURE 2.2   Cooking oil jerry cans reutilised for fetching water
                PHOTOGRAPHY BY CHRISTIANE TRISTL

Jack: It is the card who said so [general laughter]. The card said if
                 you fill your jerry can to the top, you will be charged some more
                 cents ... The normal level of the jerry can is where it should be
                 charged at three shillings. If you fill above that normal level, the
                 card tells you, you should pay more.[10]

As long as the price formula remained localised, (in)accuracies did not matter.
Everybody paid the same price per jerry can. Only with the globally connected
market device, did metrological requirements become universal. While the
price of water per litre was now conveniently traceable through the remote
monitoring system, the total price of the purchase became a tedious arithmetic operation for water users. What was the balance on the card before they
fetched water? How much was it afterwards? Sometimes there was confusion:
Is the water more expensive when it runs faster since the balance decreases
more quickly? The lifeworld of the dispenser adheres to a singular and presumably universal imagination of rationality that is in this case firmly tied to the
measurement of litres. Alternative ways of measuring water remain hidden.

5.6   Self-reliant and Individualised Human Users?
When the dispensers were implemented, smart cards with unique IDS were
distributed to every user. Smart cards enabled users to perform individualised

---

10   Interview with Jack, water dispenser user, February 2017.

transactions at the water kiosk independently of kiosk attendants. But what is a user?

Fetching water in Kondo is highly reliant on donkeys to cover the distances between the water kiosk and one's home. "Donkey power" (Kaumbutho, Waithanji, and Karimi 2004) is considered to play a significant role in Kenya's socioeconomic development. Especially in arid areas, donkeys are deliberately used for transport due to their resilience and ability to navigate the drought prone landscapes. Besides carrying goods, firewood, and people, they are also generally used for carrying water. To fetch water, two sets of two empty jerry cans are tied together with a sisal rope. The jerry cans are then lifted on the donkey's back, so that two will be dangling on the left and two on the right side of the donkey's belly (Figure 2.3). Once the jerry cans have been properly positioned, they are stabilised with another sisal rope. Once mounted, the donkey knows that it is time to fetch water and will start walking towards the kiosk. Donkeys leading the way to water kiosks, with their owners following suit, is a common sight in Kondo. Arriving at the water point, it is essential to find the hose that can be fixed to the tap of the kiosk. To fill up the jerry cans, the donkey has to be pulled close enough to the kiosk for the hose to reach. With one jerry can after the other being filled, the donkey feels the increasing weight and becomes more and more nervous. Once released, the donkey automatically starts to walk back home.

Since the dispenser has replaced the kiosk attendant, the kiosks were no longer equipped with hoses. It used to be the task of the kiosk attendant to install the hose at the tap in the morning and lock it inside the kiosk in the evening, to avoid theft. From then on, they either remained locked in the kiosk or went missing. Without the hose, it is impossible to move water from the tap to the donkey's back. The jerry cans now needed to be removed from the donkey's back to be filled up at the tap, and then lifted back up again. As Elisa, another user, explained: "I went there sometimes and then I filled the jerry cans up. And then you have to wait for a person to come by and help you. And then maybe the next person that comes is not in a position to help you. Maybe it is a child or an old person. So I stopped going there".[11] Since two of the four jerry cans are always tied together, this meant lifting two times forty kilograms back onto a mostly stubborn donkey—an impossible task for most people.

When designing the dispenser kiosk, the essential role of donkeys in retrieving water and the interplay with their human and material counterparts was

11   Interview with Elisa, February 2017.

FIGURE 2.3   Donkey in Kondo equipped for fetching water
              PHOTOGRAPHY BY CHRISTIANE TRISTL

not taken into account. The user-friendly interface had imagined "individual agents with perfectly stabilised competencies" (Callon 1998b, 8). However, in Kondo, the individual user is an assemblage of a person with a donkey companion, four jerry cans, and sisal ropes, in hope of a hose. Some users stopped going to the dispenser kiosk, while others started to organise in groups to fetch water so that they could assist each other in lifting the heavy jerry cans. Instead of individualised, self-reliant, and bounded users who inhabit the lifeworld of the dispenser as imagined by its designers, new, materially configured infrastructural collectives could be observed making their way to the water kiosks.

6   Conclusion: the Lifeworlds of Marketisation

The morality of markets invoked in recent development narratives and policies is directed towards the provision and delivery of water to "the poor." The dispenser is legitimised by the ideological belief in the functioning of flawlessly framed markets. In our case, these values are inscribed into an automated water dispenser with an intuitive interface and a mobile payment system.

The distribution of PAYGo water dispensers implicitly assumed that the waiver of meaning of the technicised lifeworld and the "no need for explanation" of the sociotechnical agencement would merge into the "invisibility of infrastructure" (Star 1999). After all, this "fad[ing]" of infrastructure "into the woodwork" (Bowker and Star 1999, 34) is the most prominent feature of modernity. As an infrastructure, the water dispenser "appears only as a relational property not as a thing stripped of use" (Star and Ruhleder 1996, 113)— it is a distributed agencement, an articulation of materialities constantly in formation.

However, while some relations demonstrate stablility and portablility, others remain missing or rupture easily when travelling from the laboratory conditions in a European engineering company to an African village in Kenya. Here, the material relations of the dispenser travel smoothly and are quickly reassembled, whereas the imagined lifeworld of the dispenser proves to be much less mobile. Moments of interruption, confusion, and misunderstanding are common and the sociotechnical agencement is brought into question. The lifeworld of the dispenser is the laboratory of neoclassical economics. The dispenser is predisposed to individual rationality and atomised calculation for the market to function in line with economists' modelling. Therefore, it is attached to values, organisations, policies, global corporations, and local communities of practice. It arranges a sociotechnical collective, populated by people, donkeys, hoses, plastic cards, metrological systems, screens linked to microchips, and other entities that carefully nudge the behaviour of individuals to rationally budget despite unstable income for the supply and demand of water. The travelling of the dispenser is not a smooth, boundary blurring "succession of displacements" (Latour 1983, 154). Rather, it is a singular event of infrastructural disruption. When the dispenser travels, the waiver of meaning is repealed. The morality embedded in the machine prioritises objectified transactions over thirst, and revenue over death. The *Urstiftungssinn* of the dispenser is marketisation and not water provision.

Bibliography

Akrich, Madeleine. 1992. "The De-Scription of Technical Objects." In *Shaping Technology/Building Society: Studies in Sociotechnical Change*, edited by Wiebe E. Bijker and John Law, 205–24. Cambridge, MA: MIT Press.

Akrich, Madeleine, and Bruno Latour. 1992. "A Summary of a Convenient Vocabulary for the Semiotics of Human and Nonhuman Assemblies." In *Shaping Technology/Building Society: Studies in Sociotechnical Change*, edited by Wiebe Bijker and John Law, 259–64. Cambridge, MA: MIT Press.

Aqua for All, Danone Communities, Stone Family Foundation, Osprey Foundation, and Hilton Foundation. 2017. "The Untapped Potential of Decentralized Solutions to Provide Safe, Sustainable Drinking Water at Large Scale: The State of the Safe Water Enterprises Market." http://safewater.enterprises/wp-content/uploads/2017/10/Highlights-V9small.pdf.

Bakker, Karen. 2008. "The Ambiguity of Community: Debating Alternatives to Private-Sector Provision of Urban Water Supply." *Water Alternatives* 1, no. 2: 236–52.

Berndt, Christian, and Marc Boeckler. 2016. "Behave, Global South! Economics, Experiments, Evidence." *Geoforum* 70: 22–24.

Berndt, Christian, and Marc Boeckler. 2017. "Economics, Experiments, Evidence: Poor Behavior and the Development of Market Subjects." In *Assembling Neoliberalism: Expertise, Practices, Subjects*, edited by Vaughan Higgins and Wendy Larner, 282–302. New York: Palgrave Macmillan.

Bishop, Matthew. 2006. "The Birth of Philanthrocapitalism." *The Economist*, February 25, 2006. www.economist.com/special-report/2006/02/25/the-birth-of-philanthrocapitalism.

Black, Maggie. 1998. "Learning What Works, 1978–1998: A 20 Year Retrospective View on International Water and Sanitation Cooperation." World Bank Water and Sanitation Program, September 30, 1998. http://documents.worldbank.org/curated/en/703661468326369198/pdf/multi-page.pdf.

Blumenberg, Hans. (1963) 2010. "Lebenswelt und Technisierung unter Aspekten der Phänomenologie." In *Hans Blumenberg. Theorie der Lebenswelt*, edited by Manfred Sommer, 181–224. Berlin: Suhrkamp.

Boeckler, Marc, and Christian Berndt. 2013. "Geographies of Circulation and Exchange III." *Progress in Human Geography* 37, no. 3: 424–32.

Bowker, Geoffrey. 1994. "Information Mythology and Infrastructure." In *Information Acumen: The Understanding and Use of Knowledge in Modern Business*, edited by Lisa Bud-Frierman, 231–47. London: Routledge.

Bowker, Geoffrey, and Susan Leigh Star. 1999. *Sorting Things Out: Classification and Its Consequences*. Cambridge, MA: MIT Press.

Brooks, Sally, Melissa Leach, Henry Lucas, and Erik Millstone. 2009. "Silver Bullets, Grand Challenges and the New Philanthropy." STEPS (Social, Technological and Environmental Pathways to Sustainability) Working Paper 24. Brighton: STEPS Centre, University of Sussex. https://steps-centre.org/anewmanifesto/manifesto_2010/clusters/cluster3/Philanthropy.pdf.

Callon, Michel. 1998a. "An Essay on Framing and Overflowing: Economic Externalities Revisited by Sociology." In *The Laws of the Markets*, edited by Michel Callon, 244–69. Oxford: Blackwell.

Callon, Michel. 1998b. "Introduction: The Embeddedness of Economic Markets in Economics." In *The Laws of the Markets*, edited by Michel Callon, 1–57. Oxford: Blackwell.

Callon, Michel, ed. 1998c. *The Laws of the Markets*. Oxford: Blackwell.

Callon, Michel. 2007. "What Does It Mean to Say That Economics Is Performative? " In *Do Economists Make Markets? On the Performativity of Economics*, edited by Donald MacKenzie, Fabian Muniesa, and Lucia Siu, 311–57. Princeton, NJ: Princeton University Press.

Callon, Michel, and Fabian Muniesa. 2005. "Peripheral Vision." *Organization Studies* 26, no. 8: 1229–50.

Cherlet, Jan. 2014. "Epistemic and Technological Determinism in Development Aid." *Science, Technology, & Human Values* 39, no. 6: 773–94.

Collier, Stephen J., Jamie Cross, Peter Redfield, and Alice Street. 2017. "Preface: Little Development Devices/Humanitarian Goods." *Limn* 9. https://limn.it/articles/precis-little-development-devices-humanitarian-goods.

Cross, Jamie. 2013. "The 100th Object: Solar Lighting Technology and Humanitarian Goods." *Journal of Material Culture* 18, no. 4: 367–87.

De Laet, Marianne, and Annemarie Mol. 2000. "The Zimbabwe Bush Pump." *Social Studies of Science* 30, no. 2: 225–63.

Fejerskov, Adam Moe. 2017. "The New Technopolitics of Development and the Global South as a Laboratory of Technological Experimentation." *Science, Technology, & Human Values* 42, no. 5: 947–68.

Ferguson, James. 2007. *Global Shadows. Africa in the Neoliberal World Order*. Durham, NC: Duke University Press.

Ferguson, James, and Akhil Gupta. 2002. "Spatializing States: Towards an Ethnography of Neoliberal Governmentality." *American Ethnologist* 29, no. 4: 981–1002.

Foster, Tim, Sean Furey, Brian Banks, and Juliet Willetts. 2019. "Functionality of Handpump Water Supplies: A Review of Data from Sub-Saharan Africa and the Asia-Pacific Region." *International Journal of Water Resources Development* 36, no. 5: 855–69.

Graham, Stephen, and Colin McFarlane, eds. 2014. *Infrastructural Lives: Infrastructure in Context*. London: Routledge.

GSMA. 2017. "Mobile for Development Utilities: Lessons from the Use of Mobile in Utility Pay-as-You-Go Models." www.gsma.com/mobilefordevelopment/wp-content/uploads/2017/01/Lessons-from-the-use-of-mobile-in-utility-pay-as-you-go-models.pdf.

GSMA. 2018. "Africa Water Enterprises: Using IoT to Monitor and Introduce Pre-Payment for Remote Water Stands in The Gambia." www.gsma.com/mobilefordevelopment /wp-content/uploads/2018/04/Africa-Water-Enterprises-Using-IoT-to-monitor -and-introduce-pre-payment-for-remote-water-stands-in-The-Gambia.pdf.

Harvey, Penny, Casper Bruun Jensen, and Atsuro Morita, eds. 2017. *Infrastructures and Social Complexity: A Companion.* London: Routledge.

Harvey, Peter, and Robert Reed. 2003. "Sustainable Rural Water Supply in Africa: Rhetoric and Reality." Conference paper presented at the 29th WEDC International Conference, Loughborough University of Technology, Abuja, Nigeria. https://repository .lboro.ac.uk/articles/conference_contribution/Sustainable_rural_water_supply_in _Africa_rhetoric_and_reality/9596414.

Ikeda, John, and Ken Liffiton. 2019. "Fintech for the Water Sector: Advancing Financial Inclusion for More Equitable Access to Water." Discussion paper. Washington, DC: World Bank. https://openknowledge.worldbank.org/entities/publication/4a197243 -4791-5d21-bfb8-3c59a0b338ad.

Irani, Lilly. 2019. *Chasing Innovation: Making Entrepreneurial Citizens in Modern India.* Princeton, NJ: Princeton University Press.

IRC (International Reference Centre for Community Water Supply and Sanitation). 1992. "International Conference on Water and the Environment: Development Issues for the 21st Century." Dublin Statement and Report of the Conference, Dublin, Ireland, January 1992. www.ircwash.org/sites/default/files/71-ICWE92-9739.pdf.

Kaerlein, Timo. 2013. "Playing with Personal Media: On an Epistemology of Ignorance." *Culture Unbound* 5: 651–70.

Kaumbutho, Pascal, Elizabeth Waithanji, and A. Karimi. 2004. "Donkey Power in the Context of Smallholder Mechanisation and Agribusiness in Kenya." In *Donkeys, People and Development: A Resource Book of the Animal Traction Network for Eastern and Southern Africa*, edited by Denis Fielding, and Paul Starkey, 94–98. Wageningen: Technical Centre for Agricultural and Rural Cooperation.

Latour, Bruno. 1983. "Give Me a Laboratory and I will Raise the World." In *Science Observed*, edited by Karin Knorr-Cetina, 141–70. London: Sage.

Latour, Bruno. 1987. *Science in Action. How to Follow Scientists and Engineers Through Society.* Cambridge, MA: Harvard University Press.

Latour, Bruno. 1991. "Technology is Society Made Durable." In *A Sociology of Monsters: Essays on Power, Technology, and Domination*, edited by John Law, 103–31. London: Routledge.

Latour, Bruno. 1992. "Where Are the Missing Masses? The Sociology of a Few Mundane Artifacts." In *Shaping Technology/Building Society: Studies in Sociotechnical Change*, edited by Wiebe E. Bijker and John Law, 225–58. Cambridge, MA: MIT Press.

Latour, Bruno. 1996. "On Actor-Network Theory: A Few Clarifications." *Soziale Welt* 47, no. 4: 369–81.

MacKenzie, Donald, Fabian Muniesa, and Lucia Siu, eds. 2007. *Do Economists Make Markets? On the Performativity of Economics*. Princeton, NJ: Princeton University Press.

McGoey, Linsey. 2016. *No Such Thing as a Free Gift. The Gates Foundation and the Price of Philanthropy*. London: Verso.

Muniesa, Fabian. 2014. *The Provoked Economy: Economic Reality and the Performative Turn*. London: Routledge.

NDMA (National Drought Management Authority). 2014. "Makueni County: Drought Monthly Bulletin for September 2014." Relief Web, October 16, 2014. https:// reliefweb.int/sites/reliefweb.int/files/resources/Makueni-September-2014.pdf.

Redfield, Peter. 2015. "Fluid Technologies: The Bush Pump, the LifeStraw and Microworlds of Humanitarian Design." *Social Studies of Science* 46, no. 2: 159–83.

Rocheleau, Dianne, Patricia Benjamin, and Alex Diang'a. 1995. "The Ukambani Region of Kenya." In *Regions at Risk: Comparisons of Threatened Environments*, edited by Jeanne X. Kasperson, Roger E. Kasperson, and B.L. Turner. Tokyo: United Nations University Press.

Roy, Ananya. 2010. *Poverty Capital: Microfinance and the Making of Development*. New York: Routledge.

Schwittay, Anke. 2014. "Designing Development: Humanitarian Design in the Financial Inclusion Assemblage." *PoLAR* 37, no. 1: 29–47.

Smits, Stef, Dick Bouman, Rozemarijn ter Horst, Arco van der Toorn, Cor Dietvorst, and Ingeborg Krukkert. 2016. "The 'End of Ownership' of Water and Sanitation Infrastructure?" Background paper to the joint IRC–VIA Water event, The Hague, The Netherlands, May 25, 2016. www.ircwash.org/resources/end-ownership-water-and -sanitation-infrastructure-background-paper-joint-irc-water-event.

Star, Susan Leigh. 1999. "The Ethnography of Infrastructure." *American Behavioral Scientist* 43, no. 3: 377–91.

Star, Susan Leigh, and Karen Ruhleder. 1996. "Steps Toward an Ecology of Infrastructure: Design and Access for Large Information Spaces" *Information Systems Research* 7, no. 1: 111–34.

Von Schnitzler, Antina. 2008. "Citizenship Prepaid: Water, Calculability, and Techno-Politics in South Africa." *Journal of Southern African Studies* 34, no. 4: 899–917.

Waldron, Daniel, Sandy Hwang, and Charles Yeboah. 2018. "Pay-as-You-Drink: Digital Finance and Smart Water Service." CGAP blog, February 13, 2018. www.cgap.org /blog/pay-you-drink-digital-finance-and-smart-water-service.

Waldron, Daniel, Caroline Frank, Akanksha Sharma, and Alexander Sotiriou. 2019. "Testing the Waters: Digital Payments for Water and Sanitation." Working paper, Washington, DC: CGAP (Consultative Group to Assist the Poor). www.cgap.org /research/publication/testing-waters-digital-payments-water-and-sanitation.

Wigdor, Daniel, and Dennis Wixon. 2011. *Brave NUI World: Designing Natural User Interfaces for Touch and Gesture.* Amsterdam: Morgan Kaufmann of Elsevier.

World Bank. 2013. "Kenya: Sub-Saharan Africa (Developing Only). World Development Indicators. Accessed April 6, 2020. https://web.archive.org/web/20131211044508/http://data.worldbank.org//country//kenya.

Zelizer, Viviana A. 1997. *The Social Meaning of Money.* Princeton, NJ: Princeton University Press.

CHAPTER 3

Crude Texting: Mobile Phones and the Infrastructuring of Protests in Oil-Age Niger

Jannik Schritt

1 Introduction

On November 28, 2011, Niger's first oil refinery was inaugurated in Zinder. What had been prepared by the recently inaugurated Nigerien President Mahamadou Issoufou as a major celebration would turn into urban youth riots, in which two people were killed, and several injured. One of the key figures in organising resistance against the new government of Issoufou was Dan Dubai,[1] a close confidant of former President Mamadou Tandja (1999–2010) who had been ousted in a military coup the year before. Dan Dubai had only returned from Dubai to Zinder in 2007 where he constructed a large villa. Talk in Zinder has it that at his arrival, he literally threw money out of the window of his Hummer SUV while driving through the streets of the city. This was not only meant to signal his arrival but also his generosity and political ambitions. Indeed, Dan Dubai is generally regarded in Niger as the founding father of *Tazarce* (meaning "continuation"), a political campaign launched in 2008 to change the constitution to abandon presidential term limits that would allow former President Tandja an unlimited number of re-elections. While the *Tazarce* campaign was initially successful, Tandja was brought down by a military coup in 2010 and Dan Dubai became part of the political opposition. Dan Dubai's civil society organisation *Mouvement Populaire pour la Pérennisation des Actions du Développement* (MPPAD) acted as one of the main architects of the riots in late 2011, organising youth in *comités de défense* (defence committees), drafting chain text messages to mobilise protesters, and spreading (mis)information. The riots were finally calmed when the military was deployed, the whole SMS network was shut down, and Niger's Prime Minister, Birgi Rafini, visited the city.

Focusing on the role of Dan Dubai and chain texting via mobile phones in the organisation of protests around the inauguration of the oil refinery, this

---

1 In Hausa Dan Dubai means "Son of Dubai"; the name can also be seen as a celebration of wealth and success associated with Dubai.

chapter analyses the interrelations of technicised interaction and direct inter-action in infrastructuring protests in Niger. Engaging with Hans Blumenberg's (1963) postulation that technology is an essential part of the lifeworld and can-not solely be understood as its coloniser—as suggested by Jürgen Habermas (2007)—this contribution looks at the role of mobile phones in Niger's life-worlds. "Lifeworlds" are thereby understood from a phenomenological stand-point as the universal structures of subjective orientations, as the world that precedes all interpretations, as given and predetermined by the sensemak-ing practices of the reflexive subject (Hitzler and Honer 1984, 58–61). "Tech-nicisation" is thereby considered an inextricable part of the lifeworld which, if successful, triggers a *Sinnentlastung* (relief from the burden of sensemak-ing) (Blumenberg 1963). While the STS literature (especially Star 1999) has emphasised that it is mainly infrastructural failures that trigger new efforts of sensemaking, I aim to show that successfully functioning infrastructures in the context of extraordinary events can have the same effect of triggering a new burden of sensemaking. The extraordinary event I focus on is an urban youth riot, and the relevant infrastructure is mobile telephony. The infrastructure was used effectively in the form of chain text messaging to coordinate the riot. This success raised new questions and disrupted the taken-for-granted signifi-cance of Niger's mobile telephony and its post-colonial political order, which together had reached a certain level of self-evidence and become the normal order of things. I will begin by looking closely at discussions about the role of information and communication technologies (ICTS) in protests, riots, and revolutions.

Since the Arab Spring, discussions about the role of ICTS in protests, riots, and revolutions resulted in numerous commentaries and articles. In a short literature overview on ICTS and social movements, Richard Heeks and Ryo Seo-Zindy (2013) show that these discussions can be grouped along two axes. Firstly, those contributions that foreground a technological determinism and explore media technologies as a central agent of social, political, and cultural change. In this sense, the Arab Spring has often been simplistically described as "Facebook" or "Twitter" revolutions. Another set of scholarly contributions follow a social deterministic approach that conceptualises, in an inverse and equally simplistic way, activist and grassroot movements as the key agents of change that use the media technologies as tools to help their cause. Secondly, contributions that follow technological determinism vary in that some pursue an optimistic and others a pessimistic viewpoint. That is, they either see media technology as a new participatory and democratising force, or they see it as a new means for exerting government control, repression, and domination (Heeks and Seo-Zindy 2013, 3–9).

Empirically, the chapter focuses on the role of Dan Dubai and so-called chain texting[2] in the protests around the opening ceremony of the oil refinery in Niger. Theoretically, the chapter starts from the proposition that as two separate fields, the juxtaposition of technoscience (operating with a means-end logic) and lifeworld (operating with a sense-making logic) is distorting (see the Introduction to this volume). It therefore follows the assumption that technology is only one of several heterogeneous elements in a material-semiotic assemblage—such as people (with skills and experience), texts (with discursively constructed meanings) and artefacts (as designed material) (Reckwitz 2002; Shove, Pantzar, and Watson 2012). It is the "distributed agency" of humans and technology (Latour 1988) which co-produces change, instead of technology being the sole agent of social, political, and cultural change. Or, to put it in the theoretical language of a "practical theory of mediation," as advanced by Antoine Hennion (2012, 259): "Highlighting the work of mediation consists of descending a little from this slightly crazy position of attributing everything to a single creator, and realising that creation is far more widely distributed, that it takes place in all the interstices between these successive mediations." And as he continues: "It is not despite the fact that there is a creator, but so that there can be a creator, that all our collective creative work is required" (Hennion 2012, 259). Thus, instead of attributing everything either to the mobile phone, or to the authors of the messages from Dan Dubai's MPPAD, I depart from an instrumentalist bias inherent in many social movement studies that focus mainly on movement leaders' strategic calculations (Ullrich and Keller 2014). I show instead that the protests emerged out of the collective creative work of human-technology interactions situated in a larger context. Moreover, analysing the protests in Niger shows how mobilisation works through a dialectic of technologically mediated interaction (chain texting) and direct face-to-face interaction (defence committees)—of *technicisation* and *lifeworlds*. Placing the human-technology interactions in a sociotechnical configuration, this contribution also follows what has been termed the "mediatisation approach" that aims at understanding the co-production of technicisation and lifeworlds by "theorising the transformation of everyday life, culture, and society in the context of the ongoing transformation of media" (Krotz 2007).

By focusing on the role of mobile phones, and in particular their function in text messaging relevant to urban protests and riots, I also aim to understand how the technology of the mobile phone—initially circulating outside Niger—is translated, malfunctions, functions differently, and becomes productive in a

---

2 "Chain texting" (sometimes also called "mass texting") meant that protesters wrote and forwarded SMSs to many receivers at once, often including all the contacts in their phone book.

distinctive sociotechnical assemblage. To empirically study the circulation of technologies across contexts, Richard Rottenburg and his colleagues suggest the concept of "travelling models" (Rottenburg 2009; Behrends, Park, and Rottenburg 2014) or "travelling technologies" (see the Introduction to this volume; see also Schnitzler 2013). Within this concept, the theory of translation takes centre stage as the underlying process that defines how technologies travel (Kaufmann and Rottenburg 2012; Freeman 2009; Schritt and Voß 2023). Here, translation refers to the process of carrying something across contexts, thus emphasising the "re-territorialisation" or "re-embedding" of travelling technologies. That is, the arriving technologies connect with existing elements at the destination, whereby not only the travelling technology itself is altered but also the whole sociotechnical assemblage into which it is integrated (a point I shall return to in the conclusion).

Focusing on translation processes of the mobile phone in Niger, I analyse the sociotechnical presuppositions of texting chain messages, their significations, as well as other dimensions of the wider sociopolitical context in Niger that were important for the protests around the inauguration of the oil refinery to materialise—namely pre-existing infrastructural networks, market contingencies, and multi-party politics. In this sense, a material-semiotic interpretation of the infrastructuring of protest reveals the entanglements of technical, economic, political, legal, and sociocultural orders with ethics, morals, affects, and imaginaries, and thus shows how pre-existing configurations are transformed in new and unpredictable ways (Calkins and Rottenburg 2017). The technology is hereby neither in and of itself liberating—i.e. as a new participatory and democratising force—nor enslaving—i.e. as a new means for government control, repression, and domination. The chapter argues that it is only by considering the larger sociotechnical configurations relevant to Niger that a deeper understanding of the role of mobile phone chain messaging in the protests surrounding the inauguration of the refinery becomes possible.[3]

I start by describing the protests with regard to the practices of chain texting. I then analyse the protest action in the context of the translation of the mobile

---

3 Empirical data was collected during 13 months of fieldwork between 2011 and 2014 using the extended case method—considered to be particularly fruitful to ethnographically analyse protest events (Schritt 2019a, 2019b). In the context of my work, this meant carrying out participant observation in civil society committees, youth groupings, and the protest events themselves, which were then situated in their larger sociotechnical context. Within this methodological framework and in addition to participant observation, I conducted over forty formal interviews and had countless informal conversations with civil society activists, youth protestors, journalists, and politicians from the government as well as the political opposition. I also tried to systematically collect media and radio coverage on the protests.

phone during political liberalisation after 1990, poor infrastructural networks and commercial contingencies, and politics from above and below. The translation focus illustrates how pre-existing infrastructural networks, capitalist contingencies, and politics in Niger's technicised lifeworlds were important for the protests around the inauguration of the oil refinery to materialise. Thus, instead of attributing the organisation of protest in an instrumentalist way to Dan Dubai and by extension, MPPAD, as the sole architects, I highlight the collective creative work of human-technology interaction.

2   Mobilising People for an Uprising

On November 28, 2011, the inauguration of Niger's first oil refinery near Zinder, the second largest city and former capital situated in the East of the country, turned violent. Two weeks before the refinery's inauguration, the government had fixed the new official Nigerien fuel price at XOF579 per litre (€0.88), well above the XOF250 per litre (€0.38) maximum that Zinder's political and social actors were demanding. Although at the time the new price was lower than the former fuel price of XOF679 (€1.04), fuel was being smuggled from Nigeria and sold in the streets of Zinder for about XOF350 per litre (€0.53).[4] Taking up the topic of oil, political actors invoked "the resource curse" to question the legitimacy of their opponents through the linguistic acts of naming, blaming, and claiming (Schritt 2020), thereby transforming oil into an idiom in which politics was framed (Schritt and Schareika 2018). Moreover, they also used the public stage of the oil refinery's inauguration to mobilise the population against the incumbent regime (Schritt 2019d). Thus, a new wave of chain messages spread rapidly among the population, especially the youth of Zinder, which forms the focus of this chapter. While the government mobilised people to attend the ceremony by paying individual activists to bring their clientele to the event to support the President, chain texting was taken up by political opponents who explicitly translated a phenomenon from the Arab Spring, which had started less than a year before in countries North of Niger. Short messages in Hausa and French, exemplified below, were initially sent from

---

4 The fuel is smuggled over the Nigerien-Nigerian border mostly in cars packed full of 50 litre jerry cans. The dealers fill the jerry cans at petrol stations and then cross the border by bypassing the border checkpoints. The street fuel price is therefore extremely volatile as it is highly dependent on subsidy policies and political events in neighbouring Nigeria. Nevertheless, even if the street fuel price is higher than at a petrol station, most Nigeriens continue to buy from roadside stalls due to personal attachments with individual street vendors, habits, mutual trust relations, and credit facilities.

unregistered SIM cards, which I explain later in the text. The political opponents questioned the government's oil policies and called on the population to
resist the government by organising themselves and boycotting the presidential arrival to the inauguration ceremony:[5]

> We do not agree—where is the quota for Zinder people in the recruit-
> ment [of oil jobs]? Where is the airport and the hospital to protect us
> from radiation and pollution [of oil]? Where are the tarmac roads, the
> congress center? We don't accept the nominations of the 9 administra-
> tors of Soraz [*Société de raffinage de Zinder*] made on an ethnocentric and
> regionalist basis.[6]
>
> SOS ghost town—let's boycott the arrival of the president by staying at
> home—let's sacrifice the day of the arrival by fasting (azoumi) to beg god
> to force our leaders to have mercy on his people.[7]

On Friday 25 November, three days before the inauguration of the oil refinery,
wealthy businessman, civil society activist, and political opponent, Dan Dubai,
was arrested for defaming the government in statements he had made during radio debates and in press releases in which he had mobilised the population against the new government of Issoufou. Talk in Zinder had it that his
arrest showed the government's fear of Dan Dubai mobilising the population
against the arrival of the President. Indeed, when the new President Mahamadou Issoufou arrived in Zinder to mark the occasion on November 28, youths
opposed his arrival. They built street barricades out of tires and set them alight,
clashed with the police, and attacked Issoufou's festive procession as it made
its way to the refinery. After the inauguration, short messages glorifying the
riots as resistance against Issoufou from the *Parti Nigerien pour la Démocratie et le Socialisme* (PNDS-Tarayya), and additionally professing faith in former
President Mamadou Tandja (1999–2010) from the *Mouvement National pour la
Société de Développement* (MNSD-Nassara), were circulated:

> Scandal in Zinder: The head of state has lost his value to Nigeriens.
> This morning, the president and the delegation that accompanied him

---

5 I have translated the mobile phone messages as closely as possible from Hausa or French.
   However, while the original messages typically contained a number of abbreviations to keep
   them short, I have written the translations in full for the sake of clarity. If there are factual
   errors or brackets, these are citations from the original messages. The text inside square
   brackets, has been inserted by me for clarity.
6 SMS received on my mobile phone during fieldwork, Zinder, November 21, 2011.
7 SMS received on my mobile phone during fieldwork, Zinder, November 28, 2011.

were made unwelcome in Zinder. The Zinderois population criticised, insulted, and threw stones at the presidential procession, destroyed official vehicles, and shouted: We want Papa Tandja back. A real sabotage of the opening ceremony for the lion.[8] Please, send this information to your brothers and sisters. It is your right. This is freedom of expression.[9]

In the days following the inauguration, more short messages aimed at mobilising protesters against the government were sent out. On December 5, the day before the trial of Dan Dubai, text messages calling on the public to attend the court case at the Zinder Tribunal were circulated:

> Democracy or dictatorship—the man of the people will be judged tomorrow Tuesday 6 December at 9 o'clock. Come to attend the trial of a man unjustly arrested by the cowards in power. The advocate of the poor, Elhadji Dan Dubai raised to the lieutenant general of the oppressed. Rendezvous not to miss—circulate the SMS.[10]

Following the call in the messages', predominantly male youths crowded in front of the court. As the trial started, Dan Dubai's supporters grew louder. The proceedings were interrupted as the crowds became entangled in battles with the police. Eventually the police began dispersing the crowd onto the streets. They were throwing stones and the police fired tear gas into the crowd. One school student who was hit by a tear gas canister died, and several others were injured. In response, the student union *Union des Scolaires Nigériens* (USN) ordered that all school and university students demonstrate against the loss of their comrade the next day. Such an order was highly effective, since school pupils usually gathered after the break for a general assembly. The union leaders expected their classmates to march collectively, creating powerful social pressure on the pupils and students to follow the USN order. In this sense, it is not only the technology of the mobile phone that must be considered in the infrastructuring of protests, but also "people as infrastructure" (Simone 2004)—without which the level of commitment and thus the success of the mobilisation could not have been explained.

On the morning of the day that followed the inauguration, pupils, students, and other youth blocked the arteries of urban Zinder life—building

---

8   *Zaki* (the lion) is a name originally given to Issoufou by his political comrades to signal his political qualities and strength but is now widely used in talking about Issoufou.

9   SMS received on my mobile phone during fieldwork, Zinder, November 28, 2011.

10   SMS received on my mobile phone during fieldwork, Zinder, December 5, 2011.

and burning barricades of tires at the squares, intersections, and main roads. During the protests, short messages continued coming through, some, such as those exemplified below, even spreading selective misinformation that added fuel to the fire:

> Late-breaking news, more than 100 police and army vehicles are on their way to Zinder. They are nearing Konni [474 km west of Zinder on the road from Niamey] and are coming to massacre the Zinderois. Circulate the SMS.[11]
>
> What will be the reaction of the Zinderois towards the killing forces that are on their way to Zinder?[12]

With both these messages powerfully reproducing a marginalised *Zinderois* identity (which I expand on later in the text) and the groundwork for peer-to-peer pressure, the crowd quickly turned into a rioting street mob, destroying traffic lights and burning down the police station at the main market. When they tried to attack the commissariat, a police officer in front of the station fired into the crowd, killing an uninvolved woman passing by. Again, that evening, more messages were sent to coordinate riots for the following day:

> A massive violent march has been organised for the whole Zinderois crowd (pupils, students, teachers, motor-taxi drivers, merchants, salesmen, workers ...) to reclaim justice for the death of two pupils targeted and killed by tear gas canisters, to reclaim justice for an innocent girl killed by a bullet from the barbaric, criminal, and murderous police. Dear brothers and sisters, come to assist this big march tomorrow at 9 am. Meeting point roundabout Total. Pass the info to the whole Zinderois population to allow them [to] mobilise and arm themselves (arrows, sling shots ...). Please circulate the message.[13]

Having placed tires and fuel at strategically important places like junctions and crossroads in the city centre during the night, male youth gangs called *palais* spearheaded the protests on the morning of December 8. Since around 2007, the phenomenon of the *palais* had emerged in urban Zinder. *Palais* are highly hierarchically organised neighbourhood groups of male youth often centred around alcohol and drug consumption with a leader referred to as

---

11   SMS received on my mobile phone during fieldwork, Zinder, December 7, 2011.
12   SMS received on my mobile phone during fieldwork, Zinder, December 7, 2011.
13   SMS received on my mobile phone during fieldwork, Zinder, December 7, 2011.

chef, boss, *président,* or *shugaba* (Souley 2012). The *palais* are dominated by a
spirit of aggressive masculinity modelled on American street gangs (Amadou
2019). Especially in marginalised city quarters in Zinder, such as Kara Kara, traditionally a neighbourhood inhabited by people affected by leprosy and today
particularly affected by poverty, high numbers of *palais* can be found. The hierarchical character of these organisations and their vertical chains of command
allow for excessive peer-to-peer pressure and thus for successful mobilisation.

Given the events of the previous days, the police were said to have been
instructed by the mayor of Zinder to stay back from the streets and only secure
strategically important points like the central police department and the
governorate. The young people were thus left unchallenged. *Palais* members
destroyed traffic lights, plundered stores and looted a bank. The bank was literally emptied and burned down, with the people taking everything they could
carry—computers, tables, chairs, even paper. Only when the military was
deployed around noon was the city finally brought under control.

As a response to the riots, the government shut down the entire SMS network in Zinder until December 11. It also used the state radio and television
channel ORTN—which has nationwide coverage—to announce a series of
initial measures, including the dismissal of executive police officers and the
convening of a commission to identify and prosecute the architects of the
riots. On the evening of December 8, after the military had restored order, the
Prime Minister, Birgi Raffini, arrived in Zinder. At the sultanate of Zinder, he
not only met with the sultan but also the governor, student representatives, the
teachers' association, the parents' association, and religious authorities. After
the meeting, a joint statement appealing for calm was released. Information
became public in the following days that the prime minister had distributed
envelopes of money to the different representatives present at the meeting.[14]
To secure peace, the military also maintained patrols of the city overnight and
throughout the following day (for a detailed description of the protests see
Schritt 2019c).

2.1 New Media and "Politics by Proxy"

Historically, strikes, demonstrations, and protests in Niger were mostly executed by the two large unions, the student union USN and the workers *Union*

---

14 The information was leaked to the public due to an internal conflict within the student
    union USN about how to distribute the money. It was reported that the religious authori-
    ties and the parent teacher association each received XOF1,000,000 (€1524), while the
    USN received XOF300,000 (€457).

des Syndicats des Travailleurs du Niger (USTN),[15] that had both formed with independence in 1960. With their short vertical chains of command, spatial proximity, and sabotage potential,—as illustrated in the case above where USN staff were able to gather pupils and students after the break—these unions are often able to organise mass joint strikes and demonstrations. This was most prominently the case on February 9, 1990, when they protested cutbacks in the public sector that had been set in Structural Adjustment Programs (SAP). Three protestors were killed and dozens were injured in these protests which proved a turning point towards the country's transition to democracy. It was with the adoption of the multiparty system in early 1991 that the state monopoly of newspapers, radio, and television was abandoned, and the freedom and independence of the media was legally established.

Since the emergence of democracy in the early 1990s, and the new and deregulated media spaces that came with it, new publics have been sprouting up in Niger. The emergence of private radio stations and mobile phones in Niger around 1997–98 saw the politicisation of issues and events in these new publics takes place "by proxy" (Kaarsholm 2009, 416), meaning that the topics become interchangeable to a certain extent. While politics were framed in terms of oil at the time of the refinery's inauguration, at other times, politics were framed in the idiom of other events such as a national holiday, the *Françafrique*, water shortages, elections, the high cost of living, or Islam (Schritt 2015). In other words, since the introduction of a multi-party system, political actors have exploited pertinent issues and events to pursue their own projects in a politically competitive environment; they have used social media, private media, and a relative freedom of speech and press to make issues public and political. To my knowledge, the protests surrounding the opening of the oil refinery in 2011 were the first in Niger in which mobilisation efforts were carried out via chain text messages.

Protest events that took place around the oil refinery's inauguration as well as thereafter show that recent politics are played out through media technologies, especially mobile phones, and radio. In the political sphere today, television and the press—which is mainly limited to the capital Niamey, with only one monthly regional newspaper in Zinder, *Le Damagaram*—play a much less significant role than mobile phones and radio. As information in Niger with its vast territory has traditionally been transmitted almost exclusively orally, illiteracy has long remained widespread and incomes low (Dan Moussa 1971), the emergence of radio as the primary medium of mass communication

---

15   Until 1978, the USTN was known as the *Union Nationale des Travailleurs du Niger* (UNTN).

is perhaps not surprising. Political, business, and civil society actors use private and associated radio stations to directly frame public debate by paying to release statements or to organise on-air debates. Nevertheless, marginalised groups such as youth, women, and rural farmers hardly have a voice in radio broadcasts. These continue to be controlled by journalists and require greater financial flow (Schritt 2020). The protests around the oil refinery's inauguration, in turn, illustrate how the mobile phone, as an instrument of political agitation, has changed the nature of the political game. Whereas social media platforms like Facebook and Twitter did not play a significant role in organising the protests—as internet access in Niger was still too limited for widespread usage at the time—simple mobile phones (not smart phones with access to Facebook and Twitter) took centre stage. Nowadays, many Nigeriens are organised in WhatsApp and Telegram groups and exchange information this way. Back then, however, chain texting allowed for the emergence of "smart mobs" (Rheingold 2002, xii) or "people who are able to act in concert even if they don't know each other." Compared to older vertical forms of union action and protests in Niger, the horizontal, decentralised, and rhizomatic nature of contemporary protests would not have been possible without new ICTS, especially mobile phones (Brunner 2017). However, as I have showed as well, the last step of mobilisation, that is going to where the action is, still partly depends on person-to-person obligations, thus making the case for a dialectic of technologically mediated interaction and direct face-to-face interaction; of technicisation and lifeworlds. I return to this issue further down in the chapter.

2.2   Poor Infrastructures and Commercial Contingencies

To avoid police persecution, the short messages, which I was sent either directly by one of the text organisers or which I collected later from youth leaders, were initially sent from unregistered SIM cards. Although SIM cards must normally be registered with the distributers when purchased, this is often not the case due to two main reasons: first, not even half of Niger's population are estimated to possesses identity cards. With births often occurring at home without any notice, registration, if at all, takes place afterwards and needs a stamp of XOF500 (€0.76) to acquire a birth certificate. An identity card is even more expensive with a collective fee and stamp of XIF2100 (€3.20) plus potential corruption money to facilitate the issuing process in a country where petty corruption is part of the everyday lifeworld (Blundo and Olivier Sardan 2006). Therefore, many Nigeriens don't have any documents to register, and the mobile phone companies would seriously limit their customer recruitment if they were to insist on a registration. Second, the mobile phone companies promote their SIM cards in rural areas where there are no electricity grids to support electronic registration. But even in urban areas, with recurrent power cuts, electronic registration can become difficult. It is also possible to circumvent SIM card registration by simply stating that one does not possess any documents. Moreover, unregistered SIM cards can also be bought informally at the market outside of distributors' shops. Taken together, the infrastructural context that exists in highly technicised countries is at least partly absent in Niger, where infrastructures of birth certificates, identity cards, and electricity grids do not have nationwide coverage. As a result, in the case of the protests, the police failed to track the source of the messages. Thus, the anonymity possible when texting via unregistered SIM cards allowed for new rhizomatic and uncontrolled forms of information dissemination, as well as organising and mobilising mass groups.

In addition, the series of chain messages were largely facilitated by a promotion that the Indian mobile phone provider Airtel had launched in Niger. The promotion gifted users who sent a message with the words "BONJOUR" to any number, with one hundred free SMS that could be sent out until midnight that same day. Established in the country only shortly before in 2011 via an acquisition of Celtel (Netherlands), and later Zain (Kuwait), Airtel had already reached a majority 68 percent market share in 2018. Generally, the Nigerien government has limited the number of telephone providers to four: Orange (France), Moov (Ivory Coast), Sahelcom (Niger state-owned, today called Niger Télécom) and Airtel (India). With a SIM card costing XOF500 (€0.76) in 2011, Nigeriens frequently held SIM cards from three, if not all four, providers. In addition to the cheap price, the widely distributed Chinese mobiles are fabricated with three plug-ins, allowing users to communicate more easily between the same provider, which is substantially cheaper than communicating between different providers. For example, when the protests unfolded around the refinery's opening, sending a short message between Airtel clients only cost XOF1 (€0.0015), but between different providers could easily have amounted to XOF10 (€0.015) or more. In sum, because of the high capital cost but very low operating costs of mobile networks, infrastructure providers have sought to attract customers with special offers that would, in turn, raise their market share. In the case of the protests, they effectively worked to not only turn mobile phones into devices of mass communication but also facilitated protest organisation. As a result, the government was forced to temporarily shut down the entire SMS network in Zinder in order to calm down the protests.

2.3   *Politics from Above and Below*

Many text messages harked back to a collective regional identity—of being *Zin-derois*—to arouse emotions of historical political marginalisation and rebel-lion. In 1898, French captain Marius Gabriel Cazemajou and his interpreter were murdered in Zinder, which had been the capital of the powerful Damaga-ram sultanate since the 18th century. The French immediately retaliated, and after facing resistance from Hausa and Tuareg warriors, ultimately defeated the Sultanate Damagaram and established the Third Military Territory of Niger in 1911, of which Zinder became the first capital. In 1926, the colonial power moved the capital to Niamey in Niger's South-West, a move which represented a major historical shift in the centre of power, from the East to the West. The French had concluded that "the Zarma" were best suited to their "civilising project" (Idrissa 2001). Since then, French colonial policy had systematically marginalised Eastern Nigeriens (especially from the perspective of the Hausa living in Zinder) in favour of western Nigeriens and the Zarma ethnic group from around the capital, Niamey. This trend continued after independence, with the Zarma constituting the country's political elite until the National Conference and the transition to democracy in the early 1990s (Ibrahim 1994).

The political marginalisation of Zinder is part of the everyday lifeworld of *Zinderois* as it is vividly evoked in present-day narratives—having contrib-uted to what some refer to as a "rebellious *Zinderois* identity" (Danda 2004). Addressing a collective ethnic Hausa or regional *Zinderois* identity has always been a strategy of the region's intellectual, political, and economic elite to acquire votes and mobilise the population, especially of the *Convention Démocratique et Sociale* (CDS-Rahama) as a traditional party of Hausaland (Zinder and Maradi) that has formed with democratisation to counter western Nigerien and Zarma ethnic hegemony in the country. Moreover, it was obvi-ous that the contents of the text messages were nearly identical to the media statements issued by political opponents. I soon learned that the texting was the coordinated action of a small group of people oscillating around the MPPAD; with one of the authors later telling me that five of them had written the texts together—backed up by the fact that some messages had identical contents but different spellings. The strategy of the political opponents was thereby two-fold: on the one hand, they wanted the wider public to refrain from attending the opening ceremony, and on the other, they were attempting to mobilise certain groups, especially *palais* (youth gangs), students, and *kabou kabou* (motorcycle taxi drivers) to stage violent protest.

During the trial of Dan Dubai, for example, one of his closer allies approached me as I stood in the crowd, proudly telling me that they had mobilised the *kabou kabou*, students, and local youth to attend. Another of his closest allies confirmed the mobilisation when I met him accidently during the riots, saying that he was working to fuel the protests. Most importantly, as a member of Dan Dubai's MPPAD that pursued a broader program of organised resistance against the new Issoufou regime, he had helped organise the urban youth of Zinder into *comités de défense,* which were attached to and given direct orders by the MPPAD. The mobilisation of youth from "above" was guaranteed both by the distribution of material rewards to their leaders and by the hierarchical structure of these groupings, which allowed youth leaders to command their followers. Dan Dubai's ultimate goal was to later turn the *comités de défense* into a political party in order to run for Niger's presidency. Moreover, Dan Dubai successfully used mobile phone messages to present himself as a "folk hero," daring to speak in the name of the poor. Mobile phone messages that were sent by his supporters and that celebrated him as a representative of the poor thereby offered a new means of public representation and helped him to acquire charismatic authority (for such a development in Uganda see Vokes 2007). Thus, in addition to the politics of new media technologies, the mobile phone messages also show the poetics of infrastructures that have diverse material qualities and cultural possibilities enabling and transforming public political life in sometimes unexpected ways (Larkin 2013).

In Niger, once you are in a powerful political or financial situation, there is significant pressure to redistribute one's wealth among social networks (Olivier de Sardan 1999). By literally throwing money out of the window of his Hummer while driving through the streets of the city, and thus distributing money outside of his personal networks, Dan Dubai had immediately signalled his political ambitions at his arrival in 2007. Due to his support for Tandja and the *Tazarce* campaign in 2008, he was able to quickly rise up the political echelons. In this political game, wealthy businessmen provide financial support for electoral campaigns, and are repaid with political postings, patronage, the embezzlement of funds, tax favours, and the distribution of public markets (Olivier de Sardan 2016). Thus, had *Tazarce* succeeded, Dan Dubai would almost certainly have been compensated with either a foothold in the Nigerien oil business or a position within the government. However, when Tandja was overthrown in a military coup in 2010, Dan Dubai was left empty-handed. One year later, in 2011, new elections were held that brought Mahamadou Issoufou's PNDS-Tarayya party into power and turned the region of Zinder into the opposition's stronghold of MNSD-Nassara and CDS-Rahama.

In general, since the emergence of a multi-party politics in the early 1990s, politics in Niger could best be described by James Scott's (1969) notion of "machine politics." For Scott, political machines are characterised by urban reward networks in which particularistic, material rewards are used to extend

control over personnel, and to maximise electoral support to produce patronage, spoils, and corruption. Party loyalty in Niger is rarely built around political
ideology, but rather around individual politicians' ability to provide material
benefits. This, in turn, produces an "activist market" with "'pop-up activists'—
that is, activists who are willing to mobilise people to attend political meetings
in return for small envelopes of cash and goodies" (McCullough, Harouna, and
Oumarou 2016, 3). As the case study of the refinery's inauguration has shown,
this underlying logic not only works for elections but also for protest mobilisation in which male youth play a central role.

Niger has the highest birth rate of any country in the world at 6.89 children born per woman, and an annual population growth of 3.3 percent. With
a median age of fifteen years, 70 percent of the population is under twenty-
five and 63 percent living below the international poverty line. Based on a
hegemonic masculinity that force young men into the role of the families'
"breadwinners," the situation for poor male youth may be described as a "culture of masculine waiting" (Masquelier 2013). Youth mass unemployment and
low-paying jobs means young men struggle to marry, have families of their
own, and many are therefore unable to see themselves, or to be seen socially,
as contributing members of the community. In a situation of "waiting", large
segments of the youth population, especially young men, organise themselves
into informal "conversation groups" as have been previously described in
this paper as *fada*[16] or *palais*. While *fada* slowly emerged in Niger during the
democratisation process in the 1990s amidst rising unemployment fuelled by
the SAP, *palais* emerged as a more recent phenomenon around 2007 in Zinder.
According to a study on youth violence in Zinder, there are approximately 320
fada or *palais*, of which 73 percent are men only, 10 percent are women only,
and 17 per cent are mixed gender groups (Souley 2012). In contrast to *fada*
that are mostly non-hierarchically organised, affirming a spirit of egalitarian-
ism and comradeship, the *palais* are highly hierarchical organisations whose
activities centre more on drug consumption, street fights, crime, and sex. Their
violence has been a matter of growing public concern for several years, especially in Zinder. In this context, the *palais* are particularly easy prey for political
machines that reward gang leaders for mobilising their followers. With high
levels of desperation and unemployment, youth are "constantly available to be
put to use for virtually any form of labour" (Hoffman 2011, 53). In other words,

---

16   *Fada* literally means "the group of people attending the judgements at the leader's pal-
      ace" (Lund 2009, 111; own translation). As the Hausa leader was traditionally the sultan,
      judgements took place in his palace. Therefore, *fada* is translated into French as *palais*.
      Nevertheless, the usage of the two terms evinces a qualitative distinction.

youth groups offer violence as a form of labour available to the highest bidder on the market, rather than as a political enactment along ideological lines (Hoffman 2011).

Nevertheless, the case study also reveals politics from below in that transmission and dissemination of the messages required youths to forward the messages to as many people as possible, as well as to consciously perform violent riots. We saw that many chain messages included a call to forward it to friends or their entire phone books. While there might have been social pressure to forward these messages, and since one's social network would be aware if a direct message had been sent to one, there is also the suspicion and danger to forward the message to supporters of the opposition (in this case the incumbent government) or even to members of the secret police. However, many youths forwarded these messages not only because of social pressure but also because they shared the widely held negative view of politicians and politics in general, and police forces in particular. With the historical sedimentation of democracy since the early 1990s, Nigeriens increasingly tend to see multiparty politics primarily as a dirty game with an infinite cycle of conflicts, rivalries, and discord (Olivier de Sardan 2017). These political games produce a negative image of the political sphere in general and have even aroused a nostalgia for Seyni Kountché's (1974–1987) military dictatorship (Olivier de Sardan 2017). In countless informal conversations I had with male youths in Niger, and in Zinder more particularly, they complained about their politicians who would not redistribute any of their wealth back to the population.

Moreover, many youths work as *kabou kabou*, as small street vendors, or are involved in smuggling fuel across the Niger-Nigerian border. These youths are deleteriously affected by police controls for motorbike documents, helmets, and contraband goods, and lament police harassment and petty corruption they endure. This situation fuels aggression against a state in general, which has lost its credibility, and against police forces in particular. The result has been an increase in youth attacks on police officers, equipment, and stations. As one of the participating youth protesters proudly said to me, their aims were to burn down the police station and kill policemen as revenge for their everyday harassment. Finally, neither having access to politics (which is controlled by a small political and economic elite), nor to the new publics that proliferated with the radio, press, and television (because these are controlled by journalist and often require financial input from the speakers), violence in post-colonies such as Niger is one of the few possibilities for youths to establish a public voice (Fanon [1961] 1988).

Having illustrated how pre-existing infrastructural networks, capitalist contingencies, and politics in Niger's technicised lifeworlds were important for the protests around the inauguration of the oil refinery to materialise, I will now conclude by returning to the questions of translation and technicisation as lifeworlds.

3   Conclusion

As Garrett (2006) has shown in an extensive literature review on social movements and ICTS, contributions can be grouped by addressing three interrelated factors: mobilising structures and organising networks, opportunity structures and the circulation of information, and framing processes and identity construction. New media technologies may thus reduce participation costs, promote collective identity, and create (networked) communities (Garrett 2006). Whether these transformations take place is, however, an empirical question: technologies should neither be understood as liberating nor subjugating in and of themselves, but rather as entangled in wider infrastructural, social, economic, and political forces which enable and restrict their potential. The case of protests around the oil refinery's inauguration in Niger illustrates how naïve observers, insensitive to the political game or the relevant empirical evidence, could easily come to celebrate new media as a democratising force. This however is an argument the empirical evidence simply does not support.

Looking at translation processes, the mobile phone is often called a "leapfrog technology" because it enables the "skipping" of particular "development stages"—in this case the development of the conventional landline telephone network which, in Western contexts, existed prior to it and in rudimentary forms, similarly in colonial Africa. The mobile phone is an exemplary case because the technology requires a relatively simple and easy to maintain infrastructure that is different from the complicated and difficult to maintain infrastructures that other travelling technologies need in order to work smoothly in Africa—such as telephone landlines, roads, railways, and electricity grids. Nevertheless, even the mobile phone does not exactly result in leapfrogging. But it can result in unexpected developments.

The case study has shown that as a travelling technology the mobile phone arrived in a technical, economic, and sociopolitical context that allowed for the anonymity of chain texting and thereby changed the nature of politics in Niger. While in highly technicised countries the registration of SIM cards is largely inevitable for the majority of users, the lack of infrastructures such as electricity grids and identity cards, as well as the predominance of everyday petty corruption, make it hard to enforce such stringent registration in Niger. The case of unregistered SIM cards shows that in a new context, where a travelling technology is not immediately nor sufficiently supported and stabilised by existing infrastructures, it will work differently than intended.

Another example of the same conclusion is provided by mobile phones fabricated in China with three plug-ins. This device allows for low-cost chain texting as messages are sent between SIM cards of the same provider and, in turn, the communication among large numbers of people becomes considerably cheaper. The Indian company Airtel established itself as a dominant player in the Nigerien telephone business in 2011 by enabling free of charge texting. In these and many similar ways, the mobile phone can open up a new scope of action that is not subjugated to the notion of leapfrogging. The new practice of chain texting among poor urban male youth gave them a voice in politics that they did not have before. The government responded by shutting down the entire SMS network, even though the mobile phone operators were among the most financially powerful companies in Niger and the mobile phone had become an important technology for all kinds of business practices. But shutting down the entire SMS network proved to be a short-term solution that only temporarily limited the newfound agency of impoverished youth.

In sum, the case study shows that the protests cannot be explained either by the technology, chain text messaging through mobile phones, nor by the human actors, the authors and senders of the messages. Rather, the case renders tangible a distributed agency between the users and their phones, and the peer-to-peer pressure that must be situated in a larger sociotechnical configuration. In Niger, during the time of my fieldwork, the technicised lifeworlds were characterised by political liberalisation after 1990, poor pre-existing infrastructures, political machines (see above), and youth gang violence. Although there were in fact at least five creators behind the chain text messages to help organise and mobilise the protests for Dan Dubai to achieve his political ambitions, these creators could exist and act only because of all the collective creative work of the entire enabling milieu (Hennion 2012).

Taking all these observations on technicisation together, the chapter explains the protests around the oil refinery's opening as the result of distributed agency. The sociopolitical lifeworld of a marginalised region had been technicised by mobile telephony to a degree that it became the natural channel through which to recruit and mobilise people for protests against the misuse of power. With the technology of the mobile phone having become part of everyday practices of the people's lifeworlds, it has—like other infrastructures, such as water and electricity supply—been drawn into processes of trivialisation and invisibilisation and in this process not only altered people's lifeworlds, but also black boxed the whole sociotechnical configuration that makes it work in the first place (Blumenberg 1963).

Finally, the infrastructuring of protest in Niger shows that it is not only breakdowns that make the workings of infrastructures visible again—an argument that has been repeatedly emphasised in the STS literature (Star 1999). In this case, it was the smooth functioning of technology that enabled the disruption of a sociopolitical figuration and at the same time made the workings of mobile telephony visible—for a while. It was the surprising success of chain messaging that disrupted the achieved renunciation of sensemaking in relation to mobile telephony and triggered a new (and probably temporary) effort at sensemaking, both in relation to the political order and to the workings of technology within it.

Acknowledgements

The research in this article was funded by the German Research Foundation (DFG) as part of the project "Oil and Social Change in Niger and Chad" (2011–2017) which was part of the DFG programme "Adaptation and Creativity in Africa" (SPP 1448). I thank the editors and the internal and external reviewers for their helpful comments. All errors remain with the author.

Bibliography

Amadou, Mamane T. 2019. "Être Héros et Parias: Les Espaces de Sociabilité des Jeunes ou Fada et Palais de Zinder." In *Identités Sahéliennes en Temps de Crise: Histoires, Enjeux et Perspectives*, edited by Baz Lecocq and Amy Niang, 319–39. Münster: LIT.

Behrends, Andrea, Sung-Joon Park, and Richard Rottenburg, eds. 2014. *Travelling Models in African Conflict Resolution: Translating Technologies of Social Ordering*. Leiden: Brill.

Blumenberg, Hans. 1963. "Lebenswelt und Technisierung unter Aspekten der Phänomenologie." *Filosofia* 14, no. 4: 855–84.

Blundo, Giorgio, and Jean-Pierre de Olivier Sardan, eds. 2006. *Everyday Corruption and the State: Citizens and Public Officials in Africa*. Cape Town, London, and New York: Zed Books.

Brunner, Elizabeth. 2017. "Wild Public Networks and Affective Movements in China: Environmental Activism, Social Media, and Protest in Maoming." *Journal of Communication* 67, no. 5: 665–77.

Calkins, Sandra, and Richard Rottenburg. 2017. "Evidence, Infrastructure and Worth." In *Infrastructures and Social Complexity: A Companion*, edited by Penelope Harvey, Casper B. Jensen, and Atsuro Morita, 253–70. London: Routledge.

Dan Moussa, Laouali. 1971. "L'information Au Niger: Mass Media, Pouvoir Publics Et Masses Populaires Dans Un Pays Du 'Tiers Monde'." Memoir presented to the IFP (Institut de Formation Professionelle).

Danda, Mahamadou. 2004. "Politique De Décentralisation, Développement Régional Et Identités Locales Au Niger : Le Cas Du Damagaram." Doctoral thesis, Institut d'études politiques de Bordeaux, Université Montesquieu – Bordeaux IV. https://halshs.archives-ouvertes.fr/tel-00370355/document.

Fanon, Frantz. (1961) 1988. *The Wretched of the Earth*. New York: Grove Press.

Freeman, Richard. 2009. "What Is 'Translation'?" *Evidence and Policy* 5, no. 4: 429–47.

Garrett, Kelly R. 2006. "Protest in an Information Society: A Review of Literature on Social Movements and New ICTs." *Information, Communication and Society* 9, no. 2: 202–24.

Habermas, Jürgen. 2007. *The Theory of Communicative Action. Volume 2: Lifeworld and System: A Critique of Functionalist Reason*. Cambridge: Polity Press.

Heeks, Richard, and Ryoung Seo-Zindy. 2013. "ICTs and Social Movements under Authoritarian Regimes: An Actor-Network Perspective." Actor-Network Theory for Development Working Paper 8. Manchester: Community Development Initiative. www.cdi.manchester.ac.uk/resources/ant4d.

Hennion, Antoine. 2012. "Music and Mediation: Towards a New Sociology of Music." In *The Cultural Study of Music: A Critical Introduction*, edited by Martin Clayton, Trevor Herbert, and Richard Middleton, 249–61. New York: Routledge.

Hitzler, Ronald, and Anne Honer. 1984. "Lebenswelt – Milieu – Situation: Terminologische Vorschläge zur Theoretischen Verständigung." *Kölner Zeitschrift für Soziologie und Sozialpsychologie* 36, no. 1: 56–74.

Hoffman, Daniel. 2011. "Violence, Just in Time: War and Work in Contemporary West Africa." *Cultural Anthropology* 26, no. 1: 34–57.

Ibrahim, Jibrin. 1994. "Political Exclusion, Democratisation and Dynamics of Ethnicity in Niger." *Africa Today* 41, no. 3: 15–39.

Idrissa, Kimba. 2001. "La Dynamique De La Gouvernance: Administration, Politique Et Ethnicité Au Niger." In *Le Niger: État Et Démocratie*, edited by Kimba Idrissa, 15–84. Paris: L'Harmattan.

Kaarsholm, Preben. 2009. "Public Spheres, Hidden Politics and Struggles over Space: Boundaries of Public Engagement in Post-Apartheid South Africa." *Social Dynamics* 35, no. 2: 411–22.

Kaufmann, Matthias, and Richard Rottenburg. 2012. "Translation als Grundoperation bei der Wanderung von Ideen." In *Kultureller Und Sprachlicher Wandel Von Wertbegriffen in Europa: Interdisziplinäre Perspektiven*, edited by Rosemarie Lühr, 219–32. Frankfurt am Main: Lang.

Krotz, Friedrich. 2007. *Mediatisierung: Fallstudien Zum Wandel Von Kommunikation*. Wiesbaden: VS Verlag für Sozialwissenschaften.

Larkin, Brian. 2013. "The Politics and Poetics of Infrastructure." *Annual Review of Anthropology*, 42, no. 1: 327–43.

Latour, Bruno. 1988. "Mixing Humans and Nonhumans Together: The Sociology of a Door-Closer." *Social Problems* 35, no. 3: 298–310.

Lund, Christian. 2009. "Les Dynamiques Politiques Locales Face À Une Démocratisation Fragile (Zinder)." In *Les Pouvoirs Locaux Au Niger*, edited by Olivier de Sardan, Jean-Pierre, and Mahaman Tidjani Alou, 89–112. Dakar: CODESRIA; Paris: Karthala.

Masquelier, Adeline. 2013. "Teatime: Boredom and the Temporalities of Young Men in Niger." *Africa* 83, no. 3: 470–91.

McCullough, Aoife, Abdoutan Harouna, and Hamani Oumarou. 2016. "The Political Economy of Voter Engagement in Niger: Research Reports and Studies." www.odi.org/sites/odi.org.uk/files/odi-assets/publications-opinion-files/10304.pdf.

Olivier de Sardan, Jean-Pierre. 1999. "A Moral Economy of Corruption in Africa?" *The Journal of Modern African Studies* 37, no. 1: 25–52.

Olivier de Sardan. 2016. "Niger: Les Quatre Prisons Du Pouvoir." *Marianne*. Accessed January 5, 2016. www.marianne.net/agora/tribunes-libres/niger-les-quatre-prisons-du-pouvoir.

Olivier de Sardan. 2017. "Rivalries of Proximity beyond the Household in Niger: Political Elites and the Baab-Izey Pattern." *Africa* 87, no. 1: 120–36.

Reckwitz Andreas. 2002. "Toward a Theory of Social Practices: A Development in Culturalist Theorizing." *European Journal of Social Theory*, 5, no. 2, 243–263.

Rheingold, Howard. 2002. *Smart Mobs: The Next Social Revolution*. Cambridge, MA: Perseus.

Rottenburg, Richard. 2009. *Far-Fetched Facts: A Parable of Development Aid*. Cambridge, MA: MIT Press.

Schnitzler, Antina von. 2013. "Traveling Technologies: Infrastructure, Ethical Regimes, and the Materiality of Politics in South Africa." *Cultural Anthropology* 28, no. 4: 670–93.

Schritt, Jannik. 2015. "The 'Protests Against Charlie Hebdo' in Niger: A Background Analysis." *Africa Spectrum* 50, no. 1: 49–64.

Schritt, Jannik. 2019a. "An Ethnography of Public Events: Reformulating the Extended Case Method in Contemporary Social Theory." *Ethnography*, November: 1–22.

Schritt, Jannik. 2019b. "Die Erweiterte Fallmethode in der Protestforschung." *Forschungsjournal Soziale Bewegungen* 32, no. 1: 58–68.

Schritt, Jannik. 2019c. "Urban Protest in Oil-Age Niger: Towards a Notion of 'Contentious Assemblages'." *Sociologus* 69, no. 1: 19–36.

Schritt, Jannik. 2019d. "Well-Oiled Protest: Adding Fuel to Political Conflicts in Niger." *African Studies Review* 62, no. 2: 49–71.

Schritt, Jannik. 2020. "Crude Talking: Radio and the Politics of Naming, Blaming and Claiming in Oil-Age Niger." *Journal of Contemporary African Studies* 38, no. 3: 415–36.

Schritt, Jannik, and Nikolaus Schareika. 2018. "Crude Moves: Oil, Power and Politics in Niger." *Africa Spectrum* 53, no. 2: 65–89.

Schritt, Jannik, and Jan-Peter Voß. 2023. "Colonization, Appropriation, Commensuration: Three Modes of Translation." *Sociological Review* https://doi.org/10.1177/00380261231201475.

Scott, James C. 1969. "Corruption, Machine Politics, and Political Change." *The American Political Science Review* 63, no. 4: 1142–58.

Shove, Elizabeth, Mika Pantzar, Matt Watson, eds. 2012. *The Dynamics of Social Practice: Everyday Life and How It Changes.* Los Angeles: Sage.

Simone, AbdouMaliq. 2004. "People as Infrastructure: Intersecting Fragments in Johannesburg." *Public Culture* 16, no. 3: 407–29.

Souley, Aboubacar. 2012. "Étude Sur Le Phénomène De Violence En Milieu Jeunes À Zinder." Niamey: UNICEF Niger.

Star, Susan Leigh. 1999. "The Ethnography of Infrastructure." *American Behavioral Scientist* 43, no. 3: 377–91.

Ullrich, Peter, and Reiner Keller. 2014. "Comparing Discourse Between Cultures." In *Conceptualizing Culture in Social Movement Research,* edited by Britta Baumgarten, Priska Daphi, and Peter Ullrich, 113–39. Basingstoke, Hants: Palgrave Macmillan.

Vokes, Richard. 2007. "Charisma, Creativity, and Cosmopolitanism: A Perspective on the Power of the New Radio Broadcasting in Uganda and Rwanda." *Journal of the Royal Anthropological Institute* 13, no. 4: 805–24.

CHAPTER 4

Between Providers and Users: Redistributors in Nairobi's Fragmented Landscape of Electricity Provision

Jonas van der Straeten and Jochen Monstadt

1   Introduction

When the United States embarked on their rural electrification programme in the 1930s, it took about two decades before 95 percent of rural settlements were electrified. In 2013, when the Kenyan electrification campaign began in earnest, the government vowed to achieve that same rate within less than half of the time. At the start of the campaign—more than a century after the first electricity generator was installed in the former colony—only a quarter of Kenyans had access to electricity (Kuo 2017). By 2020 (now revised to 2028), the government promised, Kenyans would enjoy universal access to electricity (World Bank et al. 2018; Hako 2023). The infographics and technical language of the donor brochures suggest that Kenya is currently moving faster than most other countries in sub-Saharan Africa following a seemingly linear, seamless path of modernisation and electrification. The United States (US) government-led initiative "Power Africa" awarded Kenya "best-in-class benchmark" for having achieved higher electrification rates in a short amount of time (Power Africa 2015, 17) by, for example, "rolling-out" different programmes (many of them funded by the US) to provide access to electricity in rural and informal urban areas. These programmes have not only included projects for grid expansion and intensification, but also for upscaling mini-grids and solar home systems (World Bank et al. 2018).

The metaphor of "rolling-out" reiterates an understanding of the relationship between large technical systems and society, and between system builders and users, concepts long held important to the sociology and history of technology. Drawing on historical studies in Europe and the US, scholars have conceptualised system evolution as gradually incorporating natural and social environments into the system. Hence, system building is perceived as an act of enforcing "unity from diversity, centralisation in the face of pluralism, and coherence from chaos" (Hughes 1987, 52). In this vein, electrification specifically has been described as a more-or-less deliberate attempt to align social and natural environments with a set of given technological standards and procedures in an infrastructure system. This idea has long served as a central tenet in the scholarly debate on Large Technical Systems (LTS), a field of research that scrutinises the history and dynamics of sociotechnical systems with a specific focus on infrastructures (van der Vleuten 2009). Urban infrastructure policy in East Africa, whether in the form of state developmentalist or market-based approaches, has consistently echoed this LTS narrative. The politics of urban electricity provision, for example, have long followed the ideal of a "networked city" that aims to universalise access through centralised networks, monopolist providers, and homogeneous service delivery (Monstadt and Schramm 2017).

Both the earlier LTS debate in academia and the public discourse on electricity provision in Kenya share two assumptions that pertain to the distribution of agency in LTS, and the role and everyday experience of users. First, both debates focus on centrally positioned key actors, such as utility managers and engineers, state bureaucrats, or political decision-makers who observe, articulate and—ideally—solve key problems that hamper overall system development. Their hegemonic position in the system-building (and problem-solving) process is mirrored (and justified) by the idea of technicising (see the Introduction to this volume) the everyday lives of users. The universalisation of electricity grids in Africa is widely promoted as the means to relieve users of arduous tasks like collecting firewood, operating fuel stoves, and kerosene lamps, replacing dry cell batteries or having to charge mobile phones at kiosks by introducing more stable, standardised procedures and equipment (house wiring, switches, etc.). The scenario considered ideal in the context of market economy and modernist understandings of technological progress is that of electricity networks as energy infrastructures, working silently and reliably in the background. Their everyday visibility for users is limited to making purchasing decisions and paying the bills or charging prepaid meters.

This ideal can be compromised if access to electricity is expensive and/or unreliable. Today, even in poor households in most parts of Africa, many daily practices involve the use of electricity (both on-grid and off-grid). When users struggle to pay electricity bills or prepaid credits, or experience interruptions, the infrastructure becomes painfully present. When such disruptions fade into the background and electricity-based practices no longer remind actors of the grid, it is appropriate to say that the electricity provision has been successfully technicised. The use of this non-normative term is useful because it captures an important point. If a technicised level of provision is aimed for, the utility must be attuned to the demands, financial possibilities, habits, and normative orders of its users. The taken-for-granted understanding of some basic issues in the organisation's lifeworld and in the lifeworld of the user must, at least to some extent, be congruent. This is essential because it is impossible, or at least economically unviable and socially too disruptive, to rely on controlling users who would otherwise regard electricity as free prey.

Using the empirical case of Nairobi in Kenya, the chapter unpacks the discrepancy between this ideal of highly technicised lifeworlds brought about by electricity provision and the realities of infrastructure provision on the ground. For this purpose, it explores the everyday provision of electricity in social environments with their social and technical structures that "do not lie within the control radius of the relevant dominant organisation" (Joerges 1999, 3), in this case, the national utility company. These environments, such as poor, informal urban neighbourhoods—typically framed as slums—are often characterised by limited access for state authorities. Here, electricity provision has historically depended on informal arrangements and improvised infrastructures. We argue that in these environments the utility cannot fully implement and control the technicisation process but constantly need to negotiate compromises. This is so not only with users but also with a range of actors and stakeholders who operate within the supply middle ground between utility companies and end-users. These actors, for example, landowners and landlords, community representatives, electricity co-providers and related services, as well as technicians, can therefore be defined as intermediaries, even though in most cases, they are not formally recognised as such. As Kenya's national utility has become more consolidated, the path towards the universalisation agenda has invariably required increased engagement with these actors. However, this engagement can assume very different forms, with different consequences for the lifeworld of electricity provision.

In this chapter, we scrutinise a diverse group of actors that effectively engage as (self-declared or *de facto*) intermediaries, in our case, as formal and informal redistributors of electricity from monopolist utility companies. We focus on two examples of redistributors that can be thought of as extreme within Nairobi's fragmented landscape of electricity supply. The first case is that of local cartels that redistribute electricity to slum dwellers via improvised, illegal service networks. The second is the case of the "Two Rivers" project, a land development project targeted at an upmarket clientele. One of the key selling points of this project is the creation of what Graham (2000) has termed in his work, a "premium network space". In this case study the conception includes distribution arrangements whereby a private utility—set up by the land developer—buys electricity in bulk and provides it to tenants. A backup power generator also ring-fences the local network against failures of the national grid.

Such intermediaries add a level of complexity to the social study of electricity provision in African cities which, until recently, has primarily focused on the relationship between utilities and users. While scholarship on coproviders of utility services is emerging, the highly organised redistributors under investigation here have not received much scholarly attention, with some notable exceptions (e.g., De Bercegol and Monstadt 2018). This oversight might be unsurprising, given the fact that they do not exist formally in most countries. Kenya is no exception. While generation has been unbundled and liberalised, Kenya's electricity regulation grants the utility company, Kenya Power, a monopoly as the country's sole official distribution company. Yet, redistributors matter. The case studies are just two among many forms of redistribution that are the rule rather than the exception in electricity provision in sub-Saharan Africa.

As we will show, the agency of these redistributors reaches well beyond bridging gaps in the last-mile distribution of electricity. In one way or another, they challenge what is understood to be the key strategy of system builders in the LTS debate, namely by gaining control over the intractable forces in the system environment and eliminating uncertainty (Weingart 1989). The ultimate goal of this endeavour is a state of technological closure, whereby sociotechnical relationships within the system are stabilised (i.e., no longer have to be negotiated) and regulated by standards, service contracts, and key technical components (e.g., routines to meter and sanction defaulting customers are black boxed). Notably, this concept of system building has been developed from the historical study of cases in the industrial world.

In accordance with this conventional paradigm of system building, Kenya Power has adopted an agenda which is widely promoted as the regularisation of electricity supply in urban slums and has attracted considerable support from international donors. This agenda is intended to (re)assert the hegemonic role of utilities and other formally mandated actors (e.g., regulatory authorities) and gradually eliminate informal actors that are portrayed as unruly and inefficient. However, as we highlight in this chapter, this agenda does not seem to be leading to a state of closure any time soon. The unruly redistributors in slum areas perpetually undermine this agenda by (re)negotiating the terms of supply, tinkering with institutional arrangements and technologies, and (re)opening black boxes. Moreover, they have repercussions on the national utility itself, regarding its operational strategies, self-image, professional culture, organisational culture, and its technical standards and requirements. At the same time, the emergence of a new type of formally recognised, rule-abiding redistributor in upmarket property developments has meant that the national utility can delegate responsibility for local technicisation in these cases.

This study not only addresses a wider sociological issue regarding the relationship between system building and lifeworlds in African cities, but also unpacks a challenge that becomes pertinent as infrastructural heterogeneity is increasingly recognised as a regular feature and not a deficiency of urban infrastructure in Africa: How to organise infrastructure provision outside of the realm of organisations that operate with (more) formalised rules? Utilities, regulators, and policymakers in Africa have experimented with strategies ranging from doubling down on the regularisation agenda to creating a more inclusive legal framework for formally recognised redistributors. The search for appropriate policy responses is far from over, and the details are interesting in relation to the issue of technicisation.

2   Urban Electrification in Africa

In the international development discourse, access to electricity is widely considered as a key vehicle to reduce poverty, to promote sustainable economic growth, inclusiveness, and social well-being, as well as to combat climate change, air pollution, and deforestation. Over the last few decades, governments across the African continent have adopted ambitious programmes to universalise access to electricity. At present, however, these programmes are still a long way off from achieving their target. In 2019, sub-Saharan Africa's overall electricity provision rate was only 46 percent, even though 78 percent of city dwellers had access to electricity (World Bank 2021). Rapid and unplanned urbanisation, the public utilities' limited institutional and financial resources, and the high costs of electricity are among the many obstacles to connect the remaining share of urban dwellers to the national grid (Odarno 2019). Utility companies and governments struggle to align with international standards, to push back illegal connections, to safeguard the cost recovery interests of utility companies, and to build a resilient infrastructure.

Given the demographic importance and rapid growth of "slum urbanism" (Pieterse 2011), research on urban energy provision in Africa has focused on marginalised areas where service provision is a contested issue. Much of this scholarship has revolved around the introduction of prepaid meters as a key technology to advance the agenda of regularising electricity provision in informal settlements. Studies on South Africa especially have addressed prepaid meters as a disciplining technology, designed to punish poor citizens for unruly behaviour, such as illegal connection to the grid or non-payment (Ruiters 2011), to force residents to calculate and economise their consumption, to delegate the highly contested act of disconnecting defaulting customers to an anonymous technology, and to create "spaces of calculability" (von Schnitzler 2008, 902). Other scholars take a more positive stand. In her studies on Maputo, Idalina Baptista notes that prepaid meters "facilitated the translation of everyday estimation and calculation into a desired sense of autonomy and control over electricity consumption" (2015, 1017). She conceptualises the electrification programmes in marginalised urban areas as a strategy of electricity utilities to blend various dimensions of informality in African cities into standardised routines and hegemonic practices, thus imposing rationalities on the energy system. Acknowledging that for the electricity system "to remain operational it requires constant work of maintenance and repair" (Baptista 2019, 519), the focus falls on the negotiation of access between the utility and users.

Transcending this consumption/production binary, several recent studies have revealed a multitude of actors and unpacked their role within the "heterogeneous electricity constellations" in urban Africa (Koepke 2021). Using a case study from Uganda, Paul Munro (2019) proposes the notion of "urban bricoleurs" to highlight the capacity of the urban poor to improvise everyday electricity provision from an eclectic range of sources and materials. Other case studies on East African cities have focused on neighbourhood initiatives (Andreasen and Møller-Jensen 2016); landlords or building owners who register only one electricity meter and then organise regular payments from tenants (cf. Koepke 2021; Smith 2019); or "underpaid, part-time, retrenched, or amateur technicians" who provide semi-formal (and sometimes illegal) services to electricity customers (Degani 2012, 185). Michael Degani links these intermediaries back to the political realm by positioning them in the wider arena in which electricity and its piracy has become a strategic site for Tanzanian urbanites to negotiate their social contract with the state (Degani 2023). In a study on informal settlements in Accra, Ghana, Ebenezer Amankwaa and Katherine Gough (2021) highlighted the importance of these actors in the governance of electricity provision. Most scholarship on the hybrid mixing of grid and off-grid technologies, resources, services, and actors in cities of the Global South has focused on the poor as non-networked infrastructure consumers. By including premium network spaces, this chapter takes up the call to include non-poor households and businesses in this scholarship (Lemanski 2023).

While recent scholarship has greatly advanced our understanding of intermediaries in Africa, they have barely entered scholarly debate concerned with the general characteristics and dynamics of sociotechnical transitions. This gap corresponds to a general neglect of world regions like sub-Saharan Africa or South Asia in energy related social sciences (Sovacool et al. 2020a). The debate on LTS has surprisingly little on offer to make sense of the type of actors under investigation here—even though the field has long overcome its concern with hierarchical or centralised control, has highlighted the active role of users, and now conceives system building as a "distributed, highly contested, and open-ended multi-actor game" (Sovacool et al. 2020a, 10).

In emerging studies on sociotechnical transitions through the lens of "sustainability," authors have examined different types of intermediaries who operate between consumption and production in urban infrastructure provision (Guy et al. 2011); particular groups of users who drive innovation and system change by taking on tasks of "facilitating, configuring, and brokering" (Schot, Kanger, and Verbong 2016, 3); "systemic intermediaries" which operate at system or network level (van Lente et al. 2003); or, similarly, "transitions intermediaries [...] who connect diverse groups of actors involved in transitions processes" (Sovacool et al. 2020b, 1). However, the overriding interest of these studies in innovation and system transitions determines their selection of intermediaries, narrowing the focus to those agents who are dedicated to change *by design* (e.g., NGOs, consultants, government-backed agencies, social enterprises). At the same time, recent typologies of intermediaries (e.g., in Kivimaa et al. 2019) tend to overlook those groups and individuals who act not as self-recognised but as *de facto* intermediaries with a more short-term and immediate interest in mind—such as the cartels and land-developers we describe in this case study.

The analysis of such actors contributes to an emerging research agenda that looks beyond the deficiency narratives of African cities and their alleged failure to reproduce the historical trajectories of system expansion in cities of the Global North. This research refutes the idea of "incomplete modernities" (Anderson 2002) in infrastructure provision in Africa. It takes "as a starting point not the failure of urban services and the institutions responsible for their delivery, but the vitality and multiplicity of actual delivery systems which, despite policy announcements and reforms, and notwithstanding imported models, survive and contribute to the functioning of cities" (Jaglin 2014, 434).

3   Power System Building and Reforms in Kenya: a Brief History

Throughout most of Kenya's history, the living conditions of the urban poor have proven incommensurable with the changing rationales of constructing electricity systems. Under British colonial rule, the private companies that operated under concessions from the colonial government had little incentive to expand urban provision beyond "enclave" systems (Hausman, Wilkins, and Hertner 2008, 50), supplying affluent users in the European and Asian quarters of Nairobi and Mombasa (cf. van der Straeten, forthcoming). Britain's commitment to improving welfare and infrastructure among its colonial subjects in the early 1940s had little impact on urban electricity supply. Even in the African housing estates as formalised areas of the city, access remained mostly limited to senior civil servants (van der Straeten, forthcoming). The non-networked, non-formalised parts of Nairobi remained *terra incognita* for utility managers. Attempts to develop tailored solutions for "African users" (that is the term they use in the utility) in these areas remained limited to perfunctory experiments with different technical devices such as slot meters, magnetic circuit breakers, and time switches.[1] In the late 1950s, a showcase installation in Ngecha, today a suburb of Nairobi, was described as almost grotesque by contemporary observers. In the midst of the bloody anti-Mau Mau operation, the utility wired the straw-thatched rondavel houses of the Kikuyu villagers and distributed electric razors, stoves, refrigerators, fans, suction cleaners, washing machines, and convex heaters (Hayes 1983, 313).

Kenya's political independence in 1964 hardly raised the prospects of dwellers in Nairobi's poor neighbourhoods of being connected to grid electricity. The Kenyan government adopted an agenda of nationalising and Africanising (replacing Europeans in management positions) the power utility, even though slower and less consequential than its neighbouring countries (see van der Straeten, forthcoming). In 1970, the government started acquiring about half of the utility's shares which it still holds to date (currently 50.09 percent). Under state control, planners and utility managers adopted what had become a widely accepted paradigm of top-down planning and abstract modelling that prioritised large industrial over domestic customers. Despite the introduction of cross-subsidised "lifeline tariffs" for households that consumed less than 50kWh per month, the benefits of the state-led electrification were distributed very unevenly between rural and urban areas, as well as within the cities.

In the late 1980s and early 1990s, the vertically integrated and predominantly state-owned model of electricity supply in Kenya began facing a profound crisis. Under pressure from the World Bank and other donors, the government adopted a set of measures coded in a "standard model" of power sector reform (Gratwick and Eberhard 2008). In the 1990s, the market for electricity generation was liberalised, allowing independent power producers to emerge. Generation, transmission, and distribution were unbundled, and the power utility only retained the monopoly on the latter. The government set up an independent regulatory authority and a rural electrification agency. Primarily aimed at improving the financial performance of the state-owned utilities through

---

1   See Kenya National Archives, Document KNA OP/1/757 1955, 61.

corporatisation and commercialisation (Karekezi and Kimani 2004, 13), the top-down reforms initially failed to significantly increase access to electricity among rural and urban poor (Karekezi and Kimani 2002)—as the rising number of illegal connections made evident (Turkson 2000).

By the late 2000s, however, energy sector reforms were bearing fruit. The liberalisation of the generation market attracted much private investment. The resulting increase in generation capacity helped to first mitigate and finally overcome the recurring power crises in times of drought when the state-owned hydropower plants had to be turned off (Kapika and Eberhard 2013). The power utility went from being a loss-making company to a profitable one. However, the reform and stabilisation of the sector came at the cost of affordability for electricity customers as the utility increased tariffs to a cost reflective level (Kapika and Eberhard 2013). To the present day, electricity prices in Kenya remain higher than those in much of East and Southern Africa and remain tightly protected from political intervention by an independent regulatory authority. What is generally seen as a burden for customers is widely considered a blessing by international donors. Relieved from the responsibility of increasing generation capacity, from managing recurring generation shortages, and from the arduous, unprofitable task of rural electrification, donors consider the Kenya Power utility company much better positioned today to implement new distribution projects, such as, for example, in the slum areas of Nairobi.[2]

4   The Everyday Visibility of Electricity in Nairobi's Splintered Supply Landscape

How has Kenya's eventful,—a least in the eyes of international donor agencies—yet ultimately successful history of power sector reforms impacted the everyday provision of electricity in Nairobi? Have they fulfilled the promise of invisibilising service provision that accompanies electrification programs? These questions cannot be answered without reference to the city's splintered topology that often amplifies the ambivalences and inequalities of power system construction. These inequalities can be illustrated by contrasting Nairobi's Central Business District (CBD) and the former European and Asian neighbourhoods in the Northern and Eastern part of the city with the many slum areas that have emerged since the 1950s. Today, the slums are spread over the entire

---

2   Interview with Senior Energy Specialist and Senior Urban Specialist, World Bank, Nairobi, July 20, 2017.

cityscape like an archipelago, forming clusters in the Southern part of the city and in the Mathare Valley to the North. Both types of settlement come with their own set of obstacles that have prevented electricity being rendered an invisible background condition of modern public and private life (Star 1999).

In the CBD and the affluent neighbourhoods that have for decades been almost fully covered by the electricity grid, one of the major issues has been its unreliability. As households and firms depend heavily on electricity in their daily business, the recurring power crises between the 1990s and 2010s caused great political controversy, particularly in these areas. Recurring blackouts not only led to an increase in private backup generation (Oseni 2012, 21), but also reinforced the desire of wealthy urbanites to find shortcuts to electricity's technicisation promise—like living, for example, in "premium network space" that would fend off the contingencies of electricity provision from the national utility.

In the slum areas, perhaps, electricity provision was never something that had effectively receded into the background of people's lives. For decades, Kenya Power has faced huge difficulties in extending service networks into these areas, even though most of them are located near its distribution infrastructure. The major obstacles include the dense, unplanned settlement structure and the absence of state control necessary for utility staff to protect transmission equipment, prevent illegal connections, or even enter certain neighbourhoods unharmed. In many slums, the void left by the absence of the utility was soon filled by informal infrastructure providers. Among the most notorious service providers are the electricity cartels that have illegally tapped transformers and supplied several hundred families through makeshift service lines that criss-cross the slums. In Mukuru, built on a former industrial dumping site, cartels started to steal electricity from nearby factory buildings. In Kibera, illegal resellers stepped up to deliver when the first transformers were installed nearby in the early 1980s (de Bercegol and Monstadt 2018). The power balance on the ground has long been in favour of the cartels. Unable to enforce its system rationale on the slums, the utility not only has to contend with a "laissez faire" policy but also to upgrade its technical equipment to prevent illegal redistribution loads from compromising the rest of the distribution grid.

Nevertheless, with their focus on cost recovery and the reduction of technical and non-technical losses, Kenya's ambitious reform agenda of the mid 2000s put new pressure on the utility to formalise electricity provision in the slums. Electricity poverty, low access rates, and illegal electricity tapping had frustrated Nairobi's ambitions to become a world-class metropolis. In addition, they did not fit into the image of the newly industrialising, middle-income country painted in the "Kenya Vision 2030". The universalisation and formalisation of electricity supply became one of the key national development goals. In line with Kenya's renewed efforts for slum upgrading, KPLC, Kenya Power and Lighting Company, embarked on a national slum electrification programme in 2007.[3] Despite using the conventional mix of subsidising connection costs, technically limiting use, and sanctioning illegal connections, the "Slum Connectivity Pilot 1" was a failure (de Bercegol and Monstadt 2018). Surveys in different informal areas of Nairobi, such as *Mukuru kwa Njenga* (CURI 2012, 25) or Kosovo village in Mathare (Cheseto 2013, 83), all showed that the share of residents purchasing electricity from illegal cartels remained around three quarters. Furthermore, reports have noted that the cartels' hidden service lines powering the one bedroom shacks are a major cause of electrocutions and the regular fires that ravage through the slums (McLagan, n.d.).[4] Electricity continues to be visible—at times, in the most pernicious ways.

In 2014, the utility launched a fresh attempt to expand and formalise electricity supply in the slums of Nairobi with funding from the Global Partnership on Output-Based Aid (GPOBA) and the World Bank's International Development Association. The second phase of its slum electrification programme was part of a wider initiative to electrify one million households nationwide per year. Notably, the Kenyan government opened the market to private mini-grid operators in order to support these initiatives in rural off-grid areas (Pedersen and Nygaard 2018). In urban areas, however, the monopoly of Kenya Power remains largely unchallenged.

4.1 Formalising Redistribution or "Informalising" the System? the Slum Electrification Programmes

Within the wider universalisation initiative, slum electrification programmes were particularly attractive as they promised to achieve high numbers of household connections—the key metrics for politicians and donors alike—at comparatively low costs per connection. For this reason, Kenya Power stepped up its slum electrification programme in 2013. To avoid the mistakes from the first phase, the international donors initiated a process of knowledge exchange. In South-South seminars and exchanges with operators from different countries, financed by the World Bank, Kenya Power tried to enhance its knowledge on how to counter resistance to the regularisation of electricity services.

In the first pilot programme, the utility had installed "load limiters" that automatically restrict consumption to 40kWh per month, for a flat rate of

---

3 Kenya Power and Lighting Company (KPLC) was the name of Kenya's national utility company prior to its renaming to Kenya Power in 2011.
4 Interview, Regional Office Nairobi North, Kenya Power, July 19, 2017.

about KES300 (US$ 2.9).[5] In some areas, it had also installed circuit breakers that cut off supply automatically if users connected devices requiring a current above seven amperes. Only the utility could reconnect users for a fee of KES500 (Cheseto 2013). The utility reduced connection costs to KES1.160 (approx. US$ 11.2 compared to an average connection cost of KES35.000) and took down the illegal power connections. KPLC's strategy proved ineffective, since illegal power connections were on the up again a few days later. To circumvent use restrictions, inhabitants vandalised the load limiters and refused to pay. The new network was poorly protected, easily accessible, and even facilitated an increase in illegal redistribution. Attempts to prevent illegal tapping caused local riots. KPLC technicians who were assigned to install meters or disconnect non-paying customers faced violence and were ultimately unable to enter certain slums without police protection. In turn, the cartels intimidated KPLC's customers and disconnected them from legal networks (de Bercegol and Monstadt 2018). Soon, the revenue from the newly installed meters had fallen to zero.[6]

In the second phase of its slum electrification programme, Kenya Power adjusted its approach regarding both its technical and sociomanagerial components. Its new strategy was to combine prepaid metering technology with a "community based approach" aimed at obtaining the users' support. It did so by partnering with local leaders and hiring local technicians in the process of formalising the service provision: "Even to get there, because of the political nature of the environment, we had to engage the local representatives, political representatives, engage the chief government, local level administration until we came to a point where we had a breakthrough."[7]

As a key component of slum electrification, Kenya Power adjusted its interactions with users, especially the tedious, bureaucratic, and conflict-laden work of billing, controlling use, and imposing sanctions for non-payment. Instead of load limiters and monthly flat rates, the utility now delegated the metering and disconnection of non-paying customers to prepayment meters, that were installed in large numbers. The prepayment meters have a display showing the remaining units and a keyboard. Users can recharge their electricity budget by purchasing "tokens" at nearly every neighbourhood shop and entering the code using the keyboard. This system, as the programme designers hoped, would allow users to manage their electricity budget, making basic electricity needs more affordable and accessible for the poor. Kenya Power maintained its subsidised "lifeline" tariff which applies to the first 50kWh with monthly

---

5 Interview, Regional Office Nairobi North, Kenya Power, July 19, 2017.
6 Interview, Kenya Power, July 19, 2017.
7 Interview, Regional Office Nairobi North, Kenya Power, July 19, 2017.

electricity costs of approximately KES590 or around US$5.7 for 50kWh (Otuki 2017). Moreover, the payment of the connection fee of KES1,160 is stretched over a period of twelve months as small instalments are deducted from the customers' purchase of prepaid tokens.

Although initially scripted for automated payment and less interaction with users, the prepayment meters did not relieve the utility from engaging with slum dwellers. On the contrary, the project design required a strong socio-managerial component to ensure the functionality and security of its technical system. Kenya Power set up a new marketing team that was tasked with preparing the ground for the project and getting access to informal networks and authorities in the slums. The team met with village elders and promoted the project in local *barazas* (public meetings):

> It's a challenge, but when you work with the community there, they will be able to facilitate you to move. We have cases where even the house could just be moved temporarily, so that a pole can pass. Of course, the education and them understanding that we're bringing legal power, this is going to be better than what they have been having. (...) our teams could go and attend those barazas so that they're able to also communicate. Like safety was one of our selling points. Sometimes, we go there and there's an electrocution because of the illegal way the wire has been passed through, poor wiring, poor work. Those could convince the elders in the slum, to allow the poles to be pushed there.[8]

Kenya Power devised an incentive system for their local contact persons that included fixed premiums for each connection they brokered. Other important mediators were the members of the county assembly, whose authority had been expanded as a result of devolution reforms in Kenya in 2013. While most assembly members refused to support the project, given the political risks attached to it, others saw it as an opportunity to launch their careers. One member, for example, established contacts with former cartel leaders. By integrating those leaders into the regularisation of electricity supply, Kenya Power tried to diffuse the subversive power of illegal organised redistributors.

> We didn't want at that point to recognise the cartels as such, because then they're going to be a problem. Of course, I believe that some of the people that were engaged could have had a link with the cartels.[9]

---

8   Interview, Distribution Department, Kenya Power, July 18, 2017.
9   Interview, Regional Office Nairobi North, Kenya Power, July 19, 2017.

The utility also tried to win over local residents, primarily unemployed youths, who had previously worked for the cartels. Taking advantage of the "National Youth Service" programme, Kenya Power offered them jobs, such as, for example, in the installation of new poles or in disconnection campaigns, also promising to train them as electricians and formalise their employment through permanent contracts. To liaise permanently with resident clients and serve as contact points for maintenance and repair claims, Kenya Power set up local offices in container-like depots that were permanently staffed with three workers.[10]

While the slum electrification programme was originally designed to integrate slums into the system of formalised rules, standards, and interfaces for electricity supply, it has had far-reaching impact on the utility itself. The commitment to Nairobi's slum dwellers has shaped the utility's organisational culture and day-to-day practices, as well as the profiles and skills of at least the staff involved in slum electrification. The case of slum electrification vividly illustrates the processes and sites of inscription and re-inscription, and the people involved in these efforts. The utility had to operate outside of its comfort zone of stable standardised procedures and sociotechnical relations that allow for the suspension of meaning (see the Introduction to this volume). It would instead engage with the complex and contingent social relationships of the slums. Staff managers, beyond their "official" functions, became actors in the informal networks that govern the slums. In their attempt to (re)establish the utility's hegemonic role as sole service provider in Nairobi, they had to acknowledge the power of the cartels. For some residents, Kenya Power is sometimes seen simply as one more electricity provider among others, surrounded by gangs.[11] As the utility began employing former cartel members, the inhabitants' perceived differences between the cartels and Kenya Power became increasingly blurred.

> Because now we realise that most of these informal sectors they're divided into small clans and each clan has an elder, when we work with the elders it becomes a bit easier and it becomes friendly, they become friendly and now we are able now to access. (...) So, you also (...) incorporate the head of the cartels, of course they will be aware of what we are planning to do but at least now they open the way for us. (...) We pay

10   Interview with Scheme Coordinator, Kenya Power, July 18, 2017.
11   Interview, Liaison Office for Kenya Power in Kibera, Nairobi, July 21, 2017.

them one hundred per person for an application, but that application has to be metered at the end of the day.[12]

The utility internalised the slums' political conflicts and tensions, which also had an ethnic dimension. In Kibera, for example, most of the local staff was ethnically Luo, whereas most of Kenya Power's management was from the Kikuyu and Kalenjin ethnic groups. Frustration grew among the technicians and liaison individuals recruited locally. They were threatened by the cartels and carried the immediate and sometimes physical risks of representing the utility in the slum, as Kenya Power did not fulfil its promise for permanent work contracts.[13]

To achieve its regularisation and universalisation goals, the utility had to further compromise on technical and legal procedures and standards. It discarded some of its requirements for grid connection, such as legal residential status and land tenure,[14] adequate wiring within a household or certain security standards, and diverted from security standards, for example, when it came to the positioning of the poles. Reacting to the recurring theft of copper wiring, the standard material for electricity transmission, the utility began using aluminium wiring, a low-cost material that transmits electrical current well but has a low resale value (de Bercegol and Monstadt 2017).

Kenya Power promotes the slum connectivity project as a success story, having scaled up the number of connected households from five thousand in the pilot phase to over one hundred and fifty thousand within just one year. Yet, neither the daily negotiation of terms of supply nor the resistance of the cartels have ended. While Kenya Power faces less resistance in slums like Kawangware, it still faces difficulties in Kibera.[15] Utility officials claim that only 10 percent of the 172,000 meters installed in Kibera are not vending,[16] but a local informant during a site visit in July 2017 estimated a share of 25–30 percent of the meters as "non-vending".[17] One interview partner suggested that:

> We have won the war in terms of the extent to which we have connected access to power (...). It has gone really high, but we risk losing the same

12   Interview with Ag. Regional Manager and Marketing Officer, Regional Office Nairobi North, Nairobi, July 19, 2017.
13   Interview, Liaison Office for Kenya Power in Kibera, Nairobi, July 21, 2017.
14   Interview, Distribution Department, Kenya Power, July 18, 2017.
15   Interview, Distribution Department, Kenya Power, July 18, 2017.
16   Interview, Regional Office Nairobi North, Kenya Power, July 19, 2017.
17   Interview, Liaison Office for Kenya Power in Kibera, Nairobi, July 21, 2017.

> trend basically by this tariff thing that is causing a problem which I think
> should be addressed.[18]

The "tariff thing" alludes to a design flaw in the billing model, as Kenya Power
has installed the meters with a pre-charged credit of 30kWh. As some of the
households consume extremely low amounts of electricity, they would only
need to recharge their meters after months. At this point, however, several
instalments of the connection fees have accumulated. Having used up their
pre-charged credit and facing the payment of cumulated connection fees
before being able to recharge the meter, many users return to the cartels.[19]
Others use legal electricity only for lighting or, if equipped, for television and
radio, while still using illegal electricity for other appliances that consume
more energy. Just as Kenya Power has relied on "institutional bricolage" for its
slum electrification strategy, so do the users.

The cartels have stepped up their tactics of resistance to defend their markets, for example, with more elaborate practices of interfering with the utility-
operated system (de Bercegol and Monstadt 2017). These practices include
tampering with the reputedly inaccessible, inviolable meter. The cartels have
managed to bribe technicians who set up the system for Kenya Power to find
out the technical subtleties of the meter, have made copies of the keys to the
protection boxes, and managed to make the circuit breakers inoperative. In
some cases, they have vandalised transformers to create overloads that destroy
customers' appliances, and then put the blame on Kenya Power. To date, the
utility has not been able to close and protect the key technical components of
its delivery system (de Bercegol and Monstadt 2017).

> It's a common problem even places which we will not call maybe slums
> to our standards, you'll find we have challenges of illegal connections and
> by-passing of meters. That's why even for the pre-paid meters we decided
> to install them at the pole, instead of the normal putting it at the meter at
> the house (…). It's a big challenge for us.[20]

At the same time, slum electrification has become an important asset for
Kenya Power in donor negotiations. The underlying narrative has helped to
unlock large grants that have changed the project's economic rationale for the

---

18   Interview, Regional Office Nairobi North, Kenya Power, July 19, 2017.
19   Interview, Distribution Department, Kenya Power, July 18, 2017.
20   Interview, Distribution Department, Kenya Power, July 18, 2017.

utility.[21] As Kenya Power is reimbursed at US$250 per connection,[22] the number of installed meters has become the key metrics instead of the number of kilowatt hours (kWh) sold. For this same reason, the utility does not seem to be concerned with the high percentage of non-paying meters. In March 2017, Kenya Power reported that nearly one million of its customers nationwide had not loaded their meters for months (Ngugi 2017). A study by Strathmore University, cited in Kenyan newspapers, concluded that slum dwellers in Nairobi pay around KES7 billion (US$67 million) to cartels (Omulo 2017). More recent reports suggest anecdotally that the cat and mouse game between Kenya Power, the cartels, and customers have continued to date (Langat 2019) and has largely subverted the initial successes of the community based approach—which remains well documented in the infographics of donor reports.

4.2   *Securing Power in Affluent Neighbourhoods*

While its unstable and contested nature renders electricity provision highly visible in the slums, most businesses and customers from the growing middle class are engaging less with the intricacies of electricity provision on an everyday level than they did in the past. The interaction with the utility is often reduced to recharging meters, as most post-payment meters have been replaced with prepayment meters. For large industrial and commercial customers, Kenya Power has automated the post-payment process by installing the so called "smart meters". As several thermal stations have come online over the past few years, power rationing has been virtually absent even in times of drought. Yet, power cuts still occur in Nairobi but rather as a consequence of maintenance or technical failures in the transmission networks.

The promise of making service provision (including electricity) reliable, affordable, and thereby invisible, remains a key selling point for land developers in Nairobi to justify higher rents. The case of the Two Rivers project provides a striking contrast to the slum electrification projects and hints at a further fragmentation of the city's electricity landscape. Two Rivers is a real estate development project covering an area of 102 acres situated in the city's diplomatic Blue Zone adjacent to some of its most affluent neighbourhoods, namely Runda, Girigiri, Muthaiga, and Nyali. The project is financed by a consortium of investment firms led by the Centum group (owned by billionaire Charles Kirubi). The mixed-use development encompasses the biggest shopping mall in sub-Saharan Africa (except for South Africa), medium density

21   Interview, Regional Office Nairobi North, Kenya Power, July 19, 2017.
22   Interview with Scheme Coordinator, Kenya Power, July 18, 2017.

residential homes featuring luxurious residential apartments, low-rise office blocks, and two hotels. The mall opened in 2017 and the first residential apartments were completed in 2018.

To make as many aspects of service provision as smooth and hence as invisible as possible for its tenants, the project developers have deliberately created a "premium network space"—an island of reliable, customised, and high-quality services (Graham 2000) within the city's otherwise often-crumbling public infrastructures. For that purpose, the developers have integrated Two Rivers' services with some of the well-functioning infrastructures in the neighbouring affluent residential areas, such as waste management or autonomous systems from private service providers. Roads were built in gated communities around Two Rivers with future allocation for mass transit areas or sections (Odongo 2016). The estate provides a police post, emergency services, a fire brigade, healthcare facilities, and a water purification plant for bulk water supply (Ndegwa 2016). The premium network spaces are a prime example of the interlacing of lifeworld and technicisation resulting from a directed (and well-resourced) effort.

In terms of electricity supply, the land developers have adopted a hybrid strategy. Centum has secured a license from the Kenyan Energy Regulatory Commission to generate and supply electricity, and is thus operating as a distribution company itself. It buys electricity in bulk and at high voltage (11kV) from Kenya Power, steps it down, and distributes it within its expansive estate. As a supplement and backup in case of power outages, it operates a diesel generator with a capacity of 10MW and a solar system with a peak capacity of 2MW.

The utility charges the same tariffs as Kenya Power, so it emulates the distribution infrastructure the national utility would have otherwise built. The actual value proposition of the Two River's utility is one of invisibility towards the estate's residents, who are protected from power outages and relieved of dealing with the slow, bureaucratic customer service of a large utility. Centum capitalises on the margin between the low industrial tariff at which it buys electricity, and the tariff charged the domestic households for its redistribution. For Kenya Power, the arrangement is equally attractive as it only needs to meter one distribution point and is relieved of collecting revenues from several hundred clients. Kenya's Regulation Commission also gave its consent to the arrangement.

> It is in our interest, because Kenya Power, they have PPAS [power purchase agreements] with the generators, but the same power is being sold. That means Kenya power still remains revenue neutral. It's something

you encourage (…) Kenya Power, instead of going to start collecting maybe from one thousand customers, you only see one customer and they would serve them better, they are serving like one point.[23]

However, the arrangement has implications for the overall system. By buying electricity in bulk, for example, the Two Rivers utility circumvents the cross-subsidies from the lifeline tariffs that are incorporated in the domestic tariff, rather than the industrial tariff for bulk supply. Going beyond the long-established practice of backup generation from diesel generators as a reactive measure in case of power outages, the Two Rivers utility actively engages in building a more elaborate sociotechnical system.

Two Rivers is the first case of an individual company holding a distribution license in Nairobi excepting Kenya Power. It seems to serve as a model project contributing to both the further fragmentation of the power system and the city itself.[24] Two Rivers was approved flagship project status for Vision 2030, Kenya's national long-term development blueprint. It has functioned as a model for much larger land development projects, like Tatu City in the North of Nairobi, which is served by what is advertised as "Africa's first privately-owned utilities operator" called Tatu Connect (TATU, n.d.). Besides water and sewerage supply, the utility provides electricity both through its own distribution network, including a 66/11kV substation connected to the national grid, and a 1MW solar power plant, at the same tariffs, like Kenya Power. Tatu promises to act as a buffer between their clients, the mayor and Kenya Power. In an apparent nod to Kenya Power's billing practices, Tatu furthermore announces that "you get to pay your bills to us (…) we have our meters that are very accurate so whatever you are paying is the real amount, we don't use estimations" (Kenyan Wall Street 2020, 3:09). To Kenya's big companies, these land development projects are presented as an alternative to the Central Business District with aging infrastructure that dates back to the country's colonial era.

5   Conclusion

Electricity infrastructure in Kenya has undergone a remarkable transition over the last few years. While access to formal electricity networks had primarily

---

23   Interview with Director Electricity Regulation and Senior Manager Power Systems, Energy Regulation Commission (ERC), Nairobi, July 21, 2017.

24   Interview with Director Electricity Regulation and Senior Manager Power Systems, Energy Regulation Commission (ERC), Nairobi, July 21, 2017.

been an exclusive privilege of residents in formal urban areas for almost a century, the poles of the power distribution lines are today—along with the transmission towers of mobile phone networks—often the most visible formal networked infrastructure in peripheral or informal urban areas; and increasingly so in rural areas as well. According to recent policy and Kenya Power rhetoric, this transition is the result of largely successful sectoral reforms, the financial consolidation of utility companies, the opening of markets to independent power producers, the uptake of an ambitious national agenda to universalise electricity access, and extensive funding from international donors.

However, the metrics in official reports and the users' everyday experience of service provision tell two different stories of electrification. As we have shown in this chapter, the ongoing project of universalising electricity access on a technoeconomic level has not resulted in a uniform experience of electricity provision on the household level. Quite the contrary, the expansion of the grid has created distinct lifeworlds of organising everyday electricity supply that differ wildly from neighbourhood to neighbourhood in a city like Nairobi. One of the major reasons is that the expansion of the grid outpaces the expansion of the utility's radius of hierarchical or centralised control and coordination. The geographical and social space between the utility's poles, transformers, local offices, and the users' houses oftentimes remains a contested, unregulated, and unstable territory. This territory is shaped by both established and new actors who take on various functions such as last-mile connections, redistribution, backup, repair, or governance.

This study has examined two strategies employed by the utility company Kenya Power, to advance the technicisation of electricity's "middle ground". The first is a programme to regularise electricity in slum areas, where cartels capitalise on the utility's lack of legitimacy, control, and access by illegally redistributing electricity to customers. While our research shows that the utility learned from the process, the results of the programme were ambiguous, at best. Contrary to its stated purpose, Kenya Power's slum electrification programme did not lead to depoliticised, regularised, and well-ordered infrastructure provision—which indeed rather often remained contested. To protect the technical components of the system, the utility company had to engage in extensive micro-management of social relations in the slums. This community based approach caused the utility to internalise the slum's informal relations, conflicts, and political tensions in its system design.

We have also described how a novel type of formal redistributor has emerged at the other end of Nairobi's socioeconomic spectrum, in the context of a wider attempt to create premium network spaces. These spaces are enclaves of infrastructural reliability, stability, and high-end customer services within a larger system that is still widely perceived as deficient. While this trend is too recent to draw out final conclusions, it shows that, despite the reforms of the power sector, independence from the national utility, and its grid and service infrastructure remains an attractive value proposition. This proposition is so attractive indeed, that land developers targeting business and upmarket residential clients take on the costly and complex endeavour of setting up private utilities.

On a final note, one might ask if this trend is a harbinger of what Sylvy Jaglin (2016, 183) has termed a "pragmatic turn," whereby monopolistic utilities in African cities are "acknowledging the plurality of demand and supply options, and therefore accepting the exclusionary rather than universal nature of the network, leaving space for non-conventional operators in urban governance". As our findings from Nairobi suggest, the most tangible effect of the "pragmatic turn" for now is that it allows a privileged few to buy their way out of the macro-level systemic relationships (and vulnerabilities) of monopolistic electricity provision. While the repercussions of these trends on the electricity system as a whole remain to be seen, the universalisation agenda seems fated to fostering further fragmentation of the urban electricity landscapes and the corresponding lifeworlds organising provisions. Even as the system expands, it remains characterised by a complex network of formal and informal, synergetic and parasitic, desired and undesired relationships among the utility, different intermediaries, and users.

Bibliography

Amankwaa, Ebenezer F., and Katherine V. Gough. 2021. "Everyday Contours and Politics of Infrastructure: Informal Governance of Electricity Access in Urban Ghana." *Urban Studies* 59, no. 12: 004209802110301. https://doi.org/10.1177/00420980211030155.

Anderson, Warwick. 2002. "Introduction: Postcolonial Technoscience." *Social Studies of Science* 32, no. 5–6: 643–58.

Andreasen, Manja Hoppe, and Lasse Møller-Jensen. 2016. "Beyond the Networks: Self-Help Services and Post-Settlement Network Extensions in the Periphery of Dar Es Salaam." *Habitat International* 53: 39–47. https://doi.org/10.1016/j.habitatint.2015.11.003.

Baptista, Idalina. 2015. "'We Live on Estimates': Everyday Practices of Prepaid Electricity and the Urban Condition in Maputo, Mozambique." *International Journal of Urban and Regional Research* 39, no. 5: 1004–19. https://doi.org/10.1111/1468-2427.12314.

Baptista, Idalina. 2019 "Electricity Services Always in the Making: Informality and the Work of Infrastructure Maintenance and Repair in an African City." *Urban Studies* 56, no. 3: 510–25. https://doi.org/10.1177/0042098018877692.

Cheseto, Moses. 2013. "Challenges in Planning for Electricity Infrastructure in Informal Settlements: Case of Kosovo Village, Mathare Valley, Nairobi." Master's Thesis, Department of Urban and Regional Planning, University of Nairobi.

De Bercegol, Rémi, and Jochen Monstadt. 2018. "The Kenya Slum Electrification Program: Local Politics of Electricity Networks in Kibera." *Energy Research & Social Science* 41: 249–58. https://doi.org/10.1016/j.erss.2018.04.007.

Degani, Michael. 2012. "Emergency Power: Time, Ethics, and Electricity in Postsocialist Tanzania." In *Cultures of Energy: Power, Practices, Technologies*, edited by Sarah Strauss, Stephanie Rupp, and Thomas Love, 177–92. Walnut Creek, CA: Left Coast Press.

Degani, Michael. 2023. *The City Electric: Infrastructure and Ingenuity in Postsocialist Tanzania*. Durham, NC: Duke University Press.

Graham, Stephen. 2000. "Constructing Premium Network Spaces: Reflections on Infrastructure Networks and Contemporary Urban Development." *International Journal of Urban and Regional Research* 24, no. 1: 183–200.

Gratwick, Katharine Nawaal, and Anton Eberhard. 2008. "Demise of the Standard Model for Power Sector Reform and the Emergence of Hybrid Power Markets." *Energy Policy* 36: 3948–60.

Guy, Simon, Simon Marvin, Will Medd, and Timothy Moss, eds. 2011. *Shaping Urban Infrastructures: Intermediaries and the Governance of Socio-Technical Networks*. London: Earthscan.

Hako, Nasi. 2023. "Kenya: Top Priorities to Achieve Universal Energy Access by 2028." *ESI Africa*, June 16, 2023. www.esi-africa.com/east-africa/kenya-top-priorities-to-achieve-universal-energy-access-by-2028.

Hausman, William J., Mira Wilkins, and Peter Hertner. 2008. *Global Electrification: Multinational Enterprise and International Finance in the History of Light and Power, 1878–2007*. Cambridge: Cambridge University Press.

Hayes, Charles. 1983. *Stima: An Informal History of EAP&L*. Nairobi: East African Power and Lighting Co.

Hughes, Thomas P. 1987. "The Evolution of Large Technological Systems." In *The Social Construction of Technological Systems: New Directions in the Sociology and History of Technology*, edited by Wiebe E. Bijker, Thomas P. Hughes, and Trevor J. Pinch, 51–82. Cambridge, MA: MIT Press.

Jaglin, Sylvy. 2014. "Regulating Service Delivery in Southern Cities: Rethinking Urban Heterogeneity." In *Handbook on Cities of the Global South*, edited by Susan Parnell, 434–47. New York: Routledge.

Jaglin, Sylvy. 2016 "Is the Network Challenged by the Pragmatic Turn in African Cities? Urban Transition and Hybrid Delivery Configurations." In *Beyond the Networked City: Infrastructure Reconfigurations and Urban Change in the North and South*, edited by Olivier Coutard and Jonathan Rutherford, 182–203. Abingdon, Oxon: Routledge.

Joerges, Bernward. 1999. "High Variability Discourse in the History and Sociology of Large Technical Systems." Quoted from SSOAR, Open Access Repository https://d-nb.info/1191939863/34, 1–31. Published in *The Governance of Large Technical Systems*, edited by Olivier Coutard, 259–90. London: Routledge.

Kapika, Joseph, and Anton Eberhard. 2013. *Power-Sector Reform and Regulation in Africa: Lessons from Kenya, Tanzania, Uganda, Zambia, Namibia and Ghana*. Cape Town: HSRC Press.

Karekezi, Stephen, and John Kimani. 2002. "Status of Power Sector Reform in Africa: Impact on the Poor." *Energy Policy* 30: 923–45.

Karekezi, Stephen, and John Kimani. 2004. "Have Power Sector Reforms Increased Access to Electricity among the Poor in East Africa?" *Energy for Sustainable Development* 8, no. 4: 10–25.

Kenyan Wall Street. 2020 "Tatu City Progress Since Inception." Kenyan Wall Street YouTube channel, April 7, 2020. www.youtube.com/watch?v=zVXYdk1Mjzo.

Kivimaa, Paula, Wouter Boon, Sampsa Hyysalo, and Laurens Klerkx. 2019. "Towards a Typology of Intermediaries in Sustainability Transitions: A Systematic Review and a Research Agenda." *Research Policy* 48, no. 4: 1062–75. https://doi.org/10.1016/j.respol.2018.10.006.

Koepke, Mathias, Jochen Monstadt, Francesca Pilo', and Kei Otsuki. 2021. "Rethinking Energy Transitions in Southern Cities: Urban and Infrastructural Heterogeneity in Dar Es Salaam." *Energy Research & Social Science* 74: 101937. https://doi.org/10.1016/j.erss.2021.101937.

Kuo, Lili. 2017. "Kenya's National Electrification Campaign Is Taking Less Than Half the Time It Took America." *Quartz*, January 16, 2017. https://qz.com/africa/882938/kenya-is-rolling-out-its-national-electricity-program-in-half-the-time-it-took-america.

Langat, Anthony. 2019. "In Kenya's Largest Slum, the World Bank Battles Cartels for Control of Electricity." *Devex*, February 14, 2019. www.devex.com/news/in-kenya-s-largest-slum-the-world-bank-battles-cartels-for-control-of-electricity-93878.

Lemanski, Charlotte. 2023. "Broadening the Landscape of Post-Network Cities: A Call to Research the Off-Grid Infrastructure Transitions of the Non-Poor." *Landscape Research* 48, no. 2: 174–86. https://doi.org/10.1080/01426397.2021.1972952.

McLagan, Mora. n.d. "The Mukuru Slum: A Lesson in Inequality." Oxfam. Accessed 20 April 2018. www.oxfam.org.uk/blogs/2014/05/mukuru-slum-a-lesson-in-inequality.

Monstadt, Jochen, and Sophie Schramm. 2017. "Toward the Networked City? Translating Technological Ideals and Planning Models in Water and Sanitation Systems in Dar Es Salaam." *International Journal of Urban and Regional Research* 41, no. 1: 104–25. https://doi.org/10.1111/1468-2427.12436.

Ndegwa, Ruirie. 2016. "How Shopping Complexes Are Transforming Cities and Towns in Kenya and South Africa." *Medium*, October 15, 2016. https://medium.com

/@ruiriendegwa/how-shopping-complexes-are-transforming-cities-and-towns-in
-kenya-and-south-africa-bd0bfb6fb136.

Ngugi, Brian. 2017. "Kenya Power Says 1m Customers on Prepaid Meters Are Not Paying." *Business Daily*, March 20, 2017, www.businessdailyafrica.com/corporate
/KPLC-customers-on-prepaid-meters-are-not-paying/539550–3857558-us8rnpz
/index.html.

Odarno, Lily. 2019. "Closing Sub-Saharan Africa's Electricity Access Gap: Why Cities
Must Be Part of the Solution." *World Resources Institute*, August 14, 2019. www.wri
.org/insights/closing-sub-saharan-africas-electricity-access-gap-why-cities-must
-be-part-solution.

Odongo, Dannish. 2016. "Centum's Two Rivers Gets Vision 2030 Flagship Project
Status." *Capital News*, January 27, 2016. www.capitalfm.co.ke/business/2016/01
/centums-two-rivers-gets-vision-2030-flagship-project-status.

Omulo, Collins. 2017. "Nairobi Loses Billions to Cartels Every Year, Shows Study." *Nation*,
June 28, 2020. www.nation.co.ke/news/Nairobi-slums-dwellers-lose-billions-to
-cartels/1056-3821760-qoei9tz/index.html.

Otuki, Neville. 2017. "State Subsidises 70pc of Power Bills on Low Usage." *Business
Daily*, May 23, 2017.

Oseni, Musiliu. 2012. "Power Outages and the Costs of Unsupplied Electricity: Evidence
from Backup Generation among Firms in Africa." Paper submitted as part of a PhD
thesis, University of Cambridge. https://citeseerx.ist.psu.edu/document?repid=rep1
&type=pdf&doi=ced66a3eb2d00a416501131cbbbf07483248c2f1.

Pedersen, Mathilde Brix, and Ivan Nygaard. 2018. "System Building in the Kenyan
Electrification Regime: The Case of Private Solar Mini-Grid Development." *Energy
Research & Social Science* 42, no. 2: 211–23. https://doi.org/10.1016/j.erss.2018.03.010.

Pieterse, Edgar. 2011. "Rethinking African Urbanism from the Slum." *Power* 26, no. 14.1:
40–48.

Power Africa. 2015. *Development of Kenya's Power Sector 2015–2020*. Nairobi: US Aid.

Ruiters, Greg. 2011. "Developing or Managing the Poor: The Complexities and Contradictions of Free Basic Electricity in South Africa (2000–2006)." *Africa Development*
36, no. 1: 119–142.

Schot, Johan, Laur Kanger, and Geert Verbong. 2016. "The Roles of Users in Shaping
Transitions to New Energy Systems." *Nature Energy* 1, no. 5: article 16054. https://doi
.org/10.1038/nenergy.2016.54.

Smith, Shaun. 2019. "Hybrid Networks, Everyday Life and Social Control: Electricity Access in Urban Kenya." *Urban Studies* 56, no. 6: 1250–66. https://doi
.org/10.1177/0042098018760148.

Sovacool, Benjamin K., David J. Hess, Sulfikar Amir, Frank W. Geels, Richard Hirsh,
Leandro Rodriguez Medina, Clark Miller, Carla Alvial Palavicino, Roopali Phadke,
Marianne Ryghaug, Johan Schot, Antti Silvast, Jennie Stephens, Andy Stirling, Bruno

Turnheim, Erik van der Veluten, Harro van Lente, and Steven Yearley. 2020a. "Socio-technical Agendas: Reviewing Future Directions for Energy and Climate Research." *Energy Research & Social Science* 70, no. 6: article 101617. https://doi.org/10.1016/j.erss.2020.101617.

Sovacool, Benjamin K., Bruno Turnheim, Mari Martiskainen, Donal Brown, and Paula Kivimaa. 2020b. "Guides or Gatekeepers? Incumbent-Oriented Transition Intermediaries in a Low-Carbon Era." *Energy Research & Social Science* 66: article 101490. https://doi.org/10.1016/j.erss.2020.101490.

Star, Susan Leigh. 1999. "The Ethnography of Infrastructure." *American Behavioral Scientist* 43, no. 3: 377–91.

TATU. n.d. "TATU Connect." Tatu City. Accessed January 16, 2022. www.tatucity.com/tatu-utilities.

Turkson, John. 2000. "Conclusion and Policy Summary." In *Power Sector Reform in Sub Saharan Africa*, edited by John Turkson, 204–16. London: Palgrave Macmillan.

Van der Straeten, Jonas. forthcoming. *Capital Grids: A Global History of Electricity in East Africa*. New York: Palgrave Macmillan.

Van der Vleuten, Erik. 2009. "Large Technical Systems." In *A Companion to the Philosophy of Technology*, edited by Jan-Kyrre B. Olsen, Stig A. Pedersen, and Vincent F. Hendricks, 218–23. Chichester, Sussex: Wiley-Blackwell.

Van Lente, Harro, Marko Hekkert, Rud Smits, and Bas van Waveren. 2003. "Roles of Systemic Intermediaries in Transition Processes." *International Journal of Innovation Management* 7, no. 3: 247–79. https://doi.org/10.1142/S1363919603000817

Von Schnitzler, Antina. 2008. "Citizenship Prepaid: Water, Calculability, and Techno-Politics in South Africa." *Journal of Southern African Studies* 34, no. 4: 899–917. https://doi.org/10.1080/03057070802456821.

Weingart, Peter. 1989. „Grosstechnische Systeme: Ein Paradigma der Verknüpfung von Technikentwicklung und sozialem Wandel?" In *Technik als sozialer Prozess*, edited by Peter Weingart, 174–96, Frankfurt am Main: Suhrkamp.

World Bank. 2021. "Access to Electricity, Urban (% of Urban Population): Sub-Saharan Africa." World Bank Data. Accessed January 16, 2022. https://data.worldbank.org/indicator/EG.ELC.ACCS.UR.ZS?locations=ZG.

World Bank, ESMAP (Energy Sector Management Assistance Program), GIZ (Deutsche Gesellschaft für Internationale Zusammenarbeit), and EIB (European Investment Bank). 2018. *Kenya National Electrification Strategy: Key Highlights*. Washington, DC: World Bank. https://pubdocs.worldbank.org/en/413001554284496731/Kenya-National-Electrification-Strategy-KNES-Key-Highlights-2018.pdf.

CHAPTER 5

The Measuring State: Technologies of Government in Uganda and Elsewhere

Sarah Biecker, Jude Kagoro and Klaus Schlichte

1   Introduction

Is there a relation between technology and the state? Can we think productively about questions of power, domination, government, and rule by studying how technology relates to lifeworlds? In this contribution we will argue that such reflections are particularly helpful to understand current structures of government in Uganda and the dynamics working in and through them. Additionally, we will show that Uganda, like most if not all countries, is undergoing a process of ever more internationalised politics that is partly hinged on the translation and implementation of circulating technology (Biecker and Schlichte 2021).

Studies of the social life of technology include political devices and artefacts, or if thought of the other way round, devices and artefacts as political.[1] As we want to show in this contribution, there are ongoing processes of technicisation that come to define how the social space of Uganda is ruled to function as a territory. Processes of technicisation, particularly when carried out through state agents, affect relations between public authority and private lives and are deeply political. In this chapter we will critically engage with and strengthen Hans Blumenberg's analytic concept of technicisation—a process through which technologies become invisible parts of lifeworlds (see the Introduction to this volume)—with ideas borrowed from STS. At the same time, we draw on Max Weber's political sociology to make the concept fruitful for our interest in power and domination. Weber's distinction between power (*Macht*) and its institutionalised form, domination (*Herrschaft*), depends, we argue, on

---

1   The distinction between a device and an artefact is difficult to consistently follow, as the two tend to merge in different ways. A device is usually understood as any piece of equipment, often mechanical or electrical, made for a specific purpose. An artefact is usually understood as an object made or shaped by human hands, but more often than not a device is used for this purpose. As a result, a "file" or "budget"—as understood in this chapter—may be an artefact in one situation and a device in another. To avoid misunderstanding, most people use the word technology to replace both (for example, see Akrich 1992).

processes of technicisation. Institutionalised power is mostly invisible power when rules and regulations are no longer noticed but have become part of the lifeworld. Technicisation is thus part of the institutionalisation of power which results in its transformation into domination which is taken-for-granted. Technicisation renders domination invisible. This is an open-ended process, and it is quintessentially political because it is about enforcing rules, even against resistance.

In a famous foundational article "On Some Categories of Interpretive Sociology," Weber ([1904] 2012) suggests that a fourfold distinction of social groups is helpful to understand how government and the institutionalisation of rule works: A first group of decision makers suggests or try to impose a set of rules, which a second group, usually comprising state officials and other trained, professional groups, have to carry out as part of their duties. This work usually hits resistance, since a third group, all those directly concerned or affected by the rules, will either try to manipulate or seek to evade the innovation. This struggle goes on and will end only when a fourth group, "the mass," as Weber put in in the parlance of his time, will be trained and habituated to the rules to a degree that they "(...) act in a way that corresponds closely to the averagely understood meaning [of the orders] and will act in that way, mostly without knowing anything about the purpose and meaning—or even the very existence—of the orders" (Weber [1904] 2012, 300). In short, our main argument is that the aim of government is the invisibilisation of rules and orders and that technology plays an important part in this. We base our argument on the constructivist understanding of the political form of the "state" as it has been developed by Foucault, Weber and Luhmann.

Our insights into how state organisations (like other forms of organisation) are produced—through the combination of devices and artefacts, rules and trained personnel that shape practice—are derived from our empirical material. The use of numbers in budget policy, the significance of devices and artefacts in the making of police files, and the turn to breathalysers in alcohol testing, all demonstrate this. Further, they serve to show how "technical objects (...) simultaneously embody and measure a set of relations between heterogeneous elements" (Akrich 1992, 205). All of our empirical cases show how technicisation works in practice: How technologies and lifeworlds become intertwined. We base our argument on three cases in which we can observe how Ugandan state officials and citizens deal with attempts to change rules and patterns of behaviour through measurement technologies. In this context, we look at the social and political life of three different technologies of government: central government budgets, police files, and breathalysers. By technologies of government, we mean combinations of trained actors (police officers, civil servants), devices and artefacts (budgets, files, breathalyser), and programs and rules that shape practices to constitute and maintain government activities. Not all three technologies of government have become uncontroversial and taken for granted in the same way. While police files have become part of an unquestioned infrastructure, the relevance of numerical codes in budgets as a means of controlling government itself remains contested, and the use of breathalysers to detect alcohol consumption in drunk driving has only been internalised by the police.

We will start with a brief outline of what we mean by a sociology of domination that stems from political sociology but has found its way into debates of International Relations (Laiz and Schlichte 2016).[2] The main part of our contribution presents the three case studies mentioned above. In the conclusion, we first summarise our empirical findings, propose some differentiations of the concept of technicisation, and finally suggest that our findings will help to develop further conceptualisations.

While contributing primarily to the guiding question of technicisation posed in this volume, we also hope to build bridges between rather separate fields, namely social anthropology, political sociology, and International Relations. Our argument about technologies of government could also contribute to a new way of thinking about the relations between societies, states, and international organisations, which is the conventional division of domains in the discipline of International Relations. Indeed, we argue that our approach opens up a new way of understanding how the world is governed. In this approach, formal organisations—including states and international organisations—are conceived not as collective actors separate from but intertwined with technologies of government. The two are inextricably intertwined—an understanding that we see as consistent not only with Blumenberg's philosophy of technology but also with what Luhmann, Weber, Bourdieu, and Foucault have thought about formal organisations. Finally, we hope that our observations contribute to what Tatjana Thelen has called "stategraphy"—the ethnography of statecraft—arguing that state structures are fluid and in constant flux, while each case has its own particular historicity (Thelen, Vetters, and von Benda-Beckmann 2017).

2   Technologies of Government

The concept of technologies of government that we aim to promote starts from the basic understanding of technology in much of STS, combining it with

---

2   In political science, the subject of international relations is distinguished from the academic discipline by writing the latter with capital letters and often using the acronym IR.

Blumenberg's notion of technicisation, and with Weberian and Foucauldian ideas. We follow the idea that "the study of technology itself can be transformed into a sociological tool of analysis" (Callon 2012, 77). Various understandings of technology have been proposed: namely, as referring to "physical objects or artefacts"; or to "activities or processes"; or to particular forms of "what people know and what they do" (Bijker, Hughes, and Pinch 2012, xliii). These distinctions are mostly of heuristic value, since in any given empirical case they are usually intermingled. Starting from this understanding, we focus on how technology is entangled with government, and thus with larger sociopolitical mechanisms addressed in social theory.

To this end, our analysis of technologies of government combines four core elements: a) social carriers, b) artefacts and devices (technologies), c) practices that bind things and actors together, and d) the semantics, symbolic representations and imaginaries. We argue that the combination of these four elements shapes and is shaped by an overall political goal, the respective idea of government. Accordingly, our study aims to show that these four elements can be combined in an analytically fruitful way. It brings together the technical and the social and moves "between the inside and outside of technical objects" (Akrich 1992, 206). With our analysis, the "boundary [between the inside and the outside] is turned into a line of demarcation traced, within a geography of delegation, between what is assumed by the technical object and the competences of other actants" (Akrich 1992, 206).

Let us now look at these four elements in detail before turning to our empirical observations. Social carriers of the technologies of government that we see at work in our chapter are Ugandan state agents, international staff of International Organisations (IOS), and aid agencies (the so-called expats), but also Ugandan citizens who relate in different ways to state structures and symbolic orders in their everyday actions. Theories of International Relations typically divide the world into "national ruling classes," "independent international experts," and "civil society". Sometimes the national ruling class, and more generally, the national elites are stereotyped as "corrupt," the foreign experts as "evil Western agents," and the civil society as "innocent and united in solidarity". We argue that these categories do not correspond to any kind of stereotypical evaluation, nor are they clearly demarcated, but rather overlap. We continue by contending that it is different social carrier groups that matter, and that the technologies of government permeate all layers of world society, including the Ugandan segment.

The points of entry into our case studies are a few selected technologies, all part of the formal procedures of bureaucratic organisations. We examine the workings of reports, understood as written documents that list things and thereby implicitly tell stories in order to make sense of certain aspects of the world. We focus on two versions of this form, one known as a budget and the other as a case file, both of which can be embedded in reports. Our third case study takes as its starting point the introduction of breathalysers into traffic policing in Uganda. Here we focus mainly on the negotiations between police officers and drivers being tested for alcohol consumption. It becomes clear that the introduction and use of this new device in policing only makes sense if it is accompanied by procedures that record the taken measurements so as to produce lists and reports that can then monitor the effectiveness and cost of using breathalysers, as well as their correct application of their use by police officers. Records and reports, including the equipment needed to produce them, are technologies of government, and therefore of the state. Although they are created for a specific purpose, they develop a life of their own, finding new purposes. Typically, in modern bureaucratic organisations, the handling of files or the production of registers and reports, become the actual organisational goal, while the original purpose for which all the activities are undertaken is often forgotten. In previous work, we have shown how, for example, case files are produced as some sort of lifeblood of the Ugandan Police Force (Biecker and Schlichte 2015; Biecker 2021).

We examine the selected technologies in the context of the practices to which they belong. We distinguish between practices that are associated with ideas of order and others that have other intentions, appearing to be devoid of any governmental implication. All kinds of practices that become entrenched, repeated, and institutionalised eventually become habitual routines for which the original purpose is forgotten, in other words, fades into the background. Classifying, counting, recording, documenting and listing are the core practices of government, and they all go through processes of routinisation, mostly dependent on the use of technical equipment. Essentially, they are about numerical evidence. This process is what we call technicisation. Negotiation and translation also play an important but less prominent role. It is perhaps no coincidence that classifying, counting, recording, documenting and listing are reappearing prominently in the mainstream social sciences because they have to some extent become part of government under the label of independent expertise from the academic field (Foucarde 2022). Imaginaries of a better future generates ideas about how to get there. To increase the chances of designing the right government interventions, efforts are made to measure the present, relate it to what it should be, problematise the divergence, and infer how to improve the design of the interventions to overcome the divergence. The production of this discourse involves not only bureaucrats and policy practitioners, but also the social sciences and the publics. "Policy cycles" or ideas of "good governance" are theoretical models and normative concepts that are part of such imaginaries.

Statistics (the word is related to the word state) is the main area where government and social science come together. Numbers make references. The word "reference" originates from the Latin word *referre* and means "to produce" (Latour 2002, 45). To make sense, numbers have to refer to something outside of themselves, but at the same time they need to produce something new. Numbers are the result of practices (e.g. counting, collecting, calculating) that aim to represent a given reality. Since this reality was unknown before it was captured by statistics, it is accurate to say that this reality is produced by numbers. In this sense, numbers are attempts at objectivation and officialisation (Bourdieu 2011, 206), and are thus a key part of the long-term process by which the state, as the official ordering entity, has acquired universal meaning.

Files share the basic characteristic of numbers. The German word for file, *Akte*, emphasises the quality of action, as it is derived from the Latin term *agere*, namely "to act" (Vismann 2008, 10). Both numbers and files function as referential representations of something out there, giving form to what they refer to, and certifying it as a reality—a status that those things would otherwise not hold. Police files, for example, are the result of standardised acts of measuring and writing about criminal cases, such as recording the result of a breathalyser test. In "chains of translations" (Latour 1999, 91), case files are integrated with other case files to form reports that tell a story which can eventually be used as evidence. The materialisation of this evidence can be seen in the many file folders lined up behind the prosecutor and judge in a court of law.

Both numerical statistics and files are core features of the idea of modern statehood, made universal by the process of European colonialism and subsequent postcolonial nation-building. While many other and older instances of political organisation have known forms of written documentation and information storage, the systematic transcontinental communication of knowledge in numbers and tables, in a universalised language about populations, economies, and states, is the result of colonialism. After 1945 this was primarily the work of international organisations. That the world needs to be governed and managed, and that this seems not only desirable but necessary, is a modern idea that implies technicisation. One of its current emblematic expressions is the idea of "developmentality" (Lie 2015). Officials from African governments, Western aid agencies, and international organisations all share the conviction that planning is necessary, and that it must be done with numbers, data, and reports. The idea of government became universalised as it became technicised.

The empirical part of our chapter consists of the three aforementioned case studies from Uganda. The first is on budget support, a development policy instrument aimed at improving planning that was popular among Western aid agencies in the years between the late 1990s and 2015. We show how international and Ugandan actors have adopted a particular form of numerical representation of the national government budget, even though year after year it has become clear that this attempt to manage government behaviour is not working as intended. The second case study deals with police records. Contrary to the notion of "weak states" or "failed states" in Africa, Uganda's police records have become an essential element of policing and a resource for police officers and citizens to replace complex legal procedures with out-of-court negotiations. The third study focuses on attempts by the Ugandan police to enforce drunk driving laws using recently introduced breathalysers. As we show, attempts to implement this enforcement technology have had mixed results. Police officers were quick to incorporate the new technology into their routines, however Ugandan citizens have struggled to negotiate the meaning of the machine-generated readings.

2.1 *Budget Support: Failed Attempts of Measuring the State*

Budgets presuppose planning, and budgets are plans. Uganda, like most countries, has quite a long history of both, budgets and plans. It stretches from the schemes of economic development during the British Protectorate Buganda at the beginning of the 20th century (Thompson 2003) to the first fifteen years after Uganda gained independence in 1962. Under the rule of Idi Amin (1971–1979), an informalisation of Uganda set in that endured over the years of civil war that ended in 1985 for the Southern half of the country and continued in the North for some more years. With the takeover of the current regime in 1986, "development" again became the shibboleth for far-reaching activities aiming to prepare a better future, with the joined forces of the new regime and international financial institutions, and soon, Non-Governmental Organisations (NGOs). Uganda under its new president Yoweri Museveni became a "donor darling" in the 1990s, in particular of international financial institutions. The unique situation after a long civil war and large-scale informalisation offered the chance for an adaptation to "global" blueprints of the World Bank and the International Monetary Fund.

Over the last thirty years, one scheme has followed the other. The Structural Adjustment Programs (SAP) of the 1980s overlapped with the Economy Recovery Program (ERP), which the then new government negotiated with the World Bank. Out of the latter came a Poverty Eradication Plan (PEAP), (1997–2008), a model of developmental planning for which Uganda would become a best practice scenario. This was followed in 2010 by a series of five-year National Development Plans. The next step was the realisation of the Vision 2040, which according to the newly created National Planning Authority would turn Uganda into a middle income country.[3] This plan, the primary national strategic plan, is unlike the PEAP which is less oriented towards increased social services and more towards economic growth. The current and five consecutive five-year plans were set to reach the goals of the Vision 2040. These five National Development Plans are supplemented by Sector Investment Plans (SIP) and Local Government Development Plans (LGDP), Annual Work Plans, and numerous other more down-scaled plans, reports, and assessments in the local administration.

Uganda's planning renaissance, its return to socialist technologies of government, is not a national peculiarity. As in other developing countries, it has been developed since the late 1980s with the full consent and open support, perhaps even on the initiative, of powerful external actors.[4] While the plans and visions indicate the general direction in which Uganda should be governed, the government itself has become a showcase of another political technology. This is the new instrument of development aid called General Budget Support (GBS). It has been applied in Uganda—and in at least 23 other countries like Mozambique or Nicaragua—to resist the weakening of state institutions, itself a result of bypassing the state through donors who had for a long time preferred to run projects with NGOS instead.[5]

However, its main purpose, according to its inventors and incumbent practitioners, is to make recipient states more efficient and reliable, in short, to improve governance. In practice, GBS revolves around numerical representations: specific textual genres such as tables, reports, requests, and evaluations. Its core feature is the negotiation of benchmarks in order to guide public spending, such as amounts to be allocated to certain budget items. Deviations are seen as a problem, but worse are cases where accountability is lacking, when sums disappear, and expenditure cannot be reconstructed. Budget support involves a lot of bureaucracy. It involves staff, offices and office buildings, meetings, and negotiations at all levels, coordinated action and, as we shall see,

---

3   Interview with an NPA official, Kampala, February 21, 2014.
4   Kazakhstan has a Strategiya 2030, Rwanda a Vision 2020, and Germany had an Agenda 2010.
5   Despite the image of innovation that surrounded General Budget Support, this practice is not new. Single donors, notably France in Mali, Niger, and other countries, have often directly supplemented government budgets since the 1970s. What is new is the form by which this is done.

scandals. A look at the documents produced and negotiated over the course of a financial year might give an idea of what budget support is all about.

The Joint Assessment Framework (JAF) is at the centre of negotiations between donor governments, represented by their ambassadors to Uganda and representatives of the World Bank, on the one hand, and the Ugandan government, on the other. This framework is designed and agreed upon by all donors engaged in budget support and allows them to assess the Ugandan government's measured performance equivocally. Essentially, it contains a table with benchmarks and achieved results for sectors like education, water, or mining in the budget of Uganda's central government. Officially, the benchmarks are derived from another big plan, the National Development Plan that stipulates targets in five years brackets, which, in turn are deduced from the Vision 2040 that describes the future of Uganda as a middle-income country. Some members of the "carrier group" of budgeting affirm this hierarchy of documents, and others bluntly deny any relation. A leading employee of the Department for International Development (DFID), the British development agency, who had earlier worked as a "technical advisor" in the Ugandan Office of the Prime Minister, denied the relevance of the Vision 2040 and the National Development Plan for the JAF.[6] Both documents were just created under pressure from donors and had caused considerable overlap between Ugandan agencies.

The JAF refers to the Annual Fiscal Performance Report (AFPR) which is produced twice a year by the Ugandan Ministry of Finance, Planning and Economic Development (MOFPED). Even this report, however, that mirrors the performance of Ugandan ministries, is an international product: Technical advisors of DFID and of USAID are placed within the Ministry of Finance and in the Office of the Prime Minister and assist as consultants in the production of this document.[7] The Annual Fiscal Performance Report (AFPR) compares budgeted amounts with actual spending by Uganda's eighty-three ministries and all other agencies that are part of the central government. Each edition is a detailed report of almost seven-hundred pages, broken down by sector (agriculture, land, housing, energy and mining, communication and technology, tourism and trade, education, health, water, environment, social development, security, law and order).

The AFPR is at the same time the basic source for another report, the Government Annual Performance Review (GAPR), which is produced annually after the end of each fiscal year by the Office of the Prime Minister (OPM), and which is based on additional numbers that come together at the OPM from

6 Interview Kampala, KS, February 27, 2014,
7 Interview with head of a European governmental aid agency, Kampala, KS, February 26, 2014.

all ministries. Here as well, sums budgeted in advance are compared to those actually spent. Before such a review is published, there is a "retreat" of the entire cabinet with State House officials, and the President's Office.[8] It is here, as Ugandan officials point out, that objectives, overspending, and changes are discussed, often amidst a great deal of controversy.[9] These discussions are the basis, according to Ugandan officials, upon which the governmental budget plans for the next fiscal year are negotiated between MOFPED, OPM, and the State House. The general basis for these plans is, however, the Medium-Term Budgeting Framework (MTBF), which is produced in the Office of the Prime Minister, and which lists an indexed annual value for all key titles (Lie 2015).

Apart from the international surveillance of Uganda's budget, in 2014, sixteen years after the start of GBS, attempts were made to increase parliamentarian control of the budget. A Public Finance Management Bill was passed in 2015, taking effect that same year. The bill was co-designed by experts from Norway who claim to have particular expertise in "energy governance".[10] Up until 2015, budgets were already checked by the budgetary commission of parliamentarians. There was also the Public Accounts Committee (PAC), whose members use this institution for singling-out government members for not documenting their expenses. However, in interviews, donor representatives questioned whether members of parliaments have the skills and knowledge to scrutinise government in the way that, for example, the European Parliament can. According to one interlocutor, the members of the Ugandan Parliament's Public Accounts Committee feel unable to follow all issues raised on the seven-hundred pages of the Auditor General's annual report: "They overwhelm us," he said.[11]

Although external monitoring and oversight appears to be tight, budget support has repeatedly suffered setbacks when a major embezzlement is concealed, as happened most dramatically in 2012. It is worth noting that the scandalous activities were uncovered not by donors or their staff at the highest levels of Ugandan ministries, but by the Auditor General of the Government of Uganda. This observation contradicts the general impression of Uganda as a thoroughly corrupt state, an image that also undergirds budget support as a technology of government aiming at rationalising the state. In reality, it seems

---

8   Interview with official from MOFPED Planning Unit, Kampala, KS, February 19, 2014.
9   Interview with DFID official, Kampala, KS, February 27, 2014.
10   Interview with Scandinavian diplomat, Maputo, October 4, 2012. From 2006 onwards, several oil reserves were discovered in Western Uganda. Due to immediate regulatory disputes between the Ugandan government and international oil companies, revenues are not expected before 2020.
11   Interview with former chairman of the committee, KS, March 11, 2016.

that there is quite a number of Ugandan officials that share the donor's aim of rationalising public spending.

In October 2012, the Auditor General revealed that US$ 11.6 million had disappeared from a budget intended for the rehabilitation of war-affected districts in Northern Uganda. Norway, Sweden, Denmark, and Ireland had provided substantial grants to the programme (Parliament of the Republic of Uganda 2013; Irish Aid 2014). The budget was administered by the Office of the Prime Minister (OPM). The government's Auditor General revealed in a report to Parliament that some officials in OPM, MOFPED, and the Bank of Uganda, had colluded and used "dummy accounts" to move money around so often that they believed the movements were untraceable.[12] This, as became evident later, was possible as "key controls were bypassed by the individuals who were responsible for implementing the controls" (Irish Aid 2014, 2). In interviews and informal conversations, it became clear that there is an overall agreement among Ugandan and foreign donor personnel that "corruption" is not only endemic in Uganda but is perceived as the decisive hindrance in creating more efficient state institutions. Almost all expatriates expressed that corruption Uganda's "main problem", a view shared by numerous Ugandans we interviewed.

In rationalist accounts, corruption is explained by the information problem in principal-agent relations: "Unruly agents" escape the control of superiors (Simson and Welham 2014). To us it seems less clear how to explain this pervasiveness. What is labelled "corruption" in the available sources is a term that summarises the diversion of funds, the sometimes huge gaps between scheduled funds and factual expenses, and of course, the siphoning-off of funds that just "vanish". As state officials and external observers alike confirm, all public budgets are, for example, used as funds for vote-buying in all echelons of Uganda's democracy—but there is also considerable embezzlement at work (Kjaer and Ulriksen 2020). However, there are all kinds of reasons why it is so difficult to get the actual expenses in line with earlier plans and schemes. As we have shown elsewhere with regard to the Ugandan Police Force (Biecker and Schlichte 2021), there is a huge amount of in-fighting in state agencies. Unforeseen expenses, the use of funds for political purposes, power disparities, and competition between leading personnel often leads to enormous differences between what was planned and what is done. Seemingly, budgeting as a technology of government is part of the lifeworld of state officials, but it has not taken on the form of an invisible technology. Even if budgeting itself—the need to plan expenditure—is not questioned by anybody, in practice

---

12   For details see the report of "Public Accounts Committee" of the Parliament of the Republic of Uganda (2013).

technicisation is contested. And while counting and accounting have become current practices in politics and administration, the technology does not work as it is imagined by experts, both foreign and local Ugandans.

2.2 Files: the Papery Heart of the State

In our second case, we focus on police paperwork, mainly documents and especially files, to show how they produce the state. The file is central to the technology of bureaucracy and became its core symbol, i.e. the file as the materialisation of bureaucracy.[13] In policing, the most basic file is the case file, and in Uganda, it is a collection of bundled papers tied with a piece of string.[14] The papers constitute a "chain of translation" (Latour 1999, 40), both in material terms and regarding their content. A file begins with a verbal report from a citizen who enters a police station and tells his or her story to the desk officer. The officer on duty writes a brief summary of the incident in the station book and then decides on the seriousness of the story. The officer literally translates the story into a case, and with that the citizen gets permission to go over the counter and tell the story again—this time at the Criminal Investigation Directorate (CID) desk that is a feature of every police station in the country. Here the CID officer listens to a longer and more detailed version of the complainant's oral statement and writes down the whole story. Not only the complainant, but the story as well has therefore crossed a border, the border from being just a story to being a police case, documented and materialised in ink on paper. This paper of "first information" is then transferred to the records officer, who includes it into a crime report book. The paper is given a case number, which an officer writes on a piece of paper and hands over to the waiting complainant who is then asked to follow up on the case in due course. The case itself is now handed over to an officer in charge of the investigation unit. As the investigation progresses, the file continues to build up. Everything is documented following a particular structure: the complainant's statement, the witness's statement, and the suspect's statement—if all these people are available. Finally, an officer writes a concluding statement. Depending on the case, the file may also contain other documents, medical reports, or drawings of crime scenes. It normally takes weeks, months, or even years for all these documents to be assembled and the file becomes the "graphic embodiment of an issue" (Hull 2003, 302).

---

13   At the time of the fieldwork, the documents were literally paperwork, as they had not yet been digitised.

14   For primary sources on this case see Biecker (2021).

Although few files find their way into court cases and many seem to be forgotten on the shelves of station archives, files have "data-producing power" (Popitz 1992, 32). Their very existence is a resource of power, and the acceptance of their form is constitutive of state rule. Yet files have at least two powerful roles within the Ugandan police: first, they are a tool of empowerment for citizens since they can be used effectively in out-of-court negotiations; second, they are self-referential for the police themselves. Citizens accept the files as a form by using them to informally negotiate with the conflicting party, with or without the assistance of the police. Returning from the police with a case number in hand, can be a powerful tool to start negotiations. In many criminal cases, from theft to road accidents and rape, most people in Uganda prefer compensation to long and often futile court cases. It is also common in Uganda for complainants to bring suspects and witnesses to the police station so that investigators can take their statements. Against this background, citizens often try to avoid further police involvement and use the file for their own purposes, such as threats or compensation claims. Even in this out-of-court context, the file is a document that is "immutable, presentable, readable, and combinable" and can mobilise networks of ideas and people (Latour 1988, 26). The file thus demonstrates the dialectic between the visible and invisible sides of a technology in use.

The other dimension of the power of files is their self-referentiality. Files are self-referential in the sense that they provide evidence that officers, their organisational units, and the police themselves have performed their duties correctly and on time. Since files also contain minutes, which usually forms the last part of the entire document, they are a "written manifestation of an act" (Vismann 2000, 87). These minutes, written by the responsible officer and her superior, is understood as a "diary of investigation" that documents every order and investigation. In this regard, files are their own protocols (Vismann 2000, 87) and a "communicative practice fundamentally organised by 'graphic artefacts'" (Hull 2003, 287). Since in our case study, superiors and officers often share one office or even desk, files are simply moved between one desk or office for signatures and orders. The ordered procedure of file-keeping creates a "metatext which emplots the official career of the artefact in time, space, organisational order, and the order of the graphic artefacts" (Hull 2003, 297). And at the same time the files form a "chain" (Riles 2006, 80) which in some cases is continued in a court of law.

So far, our study of how written words and numbers create budget plans and police files confirms Matthew Hull's statement: "Writing establishes the stable relation between words and things necessary for bureaucracies" (Hull 2003, 291). In this context of bureaucratic government, the technology of writing also establishes a relation between words and practices as the technology of filing triggers further technologies and practices, for example the case file leads to the investigation of the case. More importantly, our investigation has brought a new aspect to light. On the one hand, files are part of institutionalised power, as explained in the introduction of this chapter. The police operate as the "decision makers" in Weber's terminology, and the citizens—as the "mass"—are supposed to follow. And indeed, as soon as a citizen voluntarily enters a police station, the logics of recording is approved and established by this very act as the unquestionable way of "doing" policing. In Uganda, neither the police nor the citizens question the act of recording and documenting and, in most cases, both sides actually demand this documentary work. Officers need the file as proof of their work and citizens need the file or the file number in order to present it to other bureaucracies, e.g. in order to get a new identity document.

More importantly, we were able to show that while the use of records is unquestioned, their purpose is not necessarily clearly defined. On the contrary, people can use police files for their own purposes, for example to negotiate with the conflicting party. They can even decide whether and how to follow up on the file which is a problem for police officers because they depend on the statements in order to file a case and thus be able to investigate. At the same time, police officers can make files disappear after payments have been made, or decide whether a file needs to be opened or not. Like our first case study, our second case study also has shown that once a technology of government has become a self-evident and unquestioned part of an organisation's lifeworld, practices that undermine its purpose are still possible. But these do not call into question the systemic nature of the records that most actors use instrumentally. Our third case study offers another possibility.

2.3   *Drunk Driving Operations*

In a particular practice of the Ugandan Police Force, another observation on technicisation can be made. In "Drunk Driving Operations," breathalysers have been appropriated and integrated into the police routines quickly and smoothly. But while for the officers the new device has become an unquestioned part of their professional lifeworld, Ugandan citizens struggle with the new technology of measurement, which they see as disrupting their own lifeworlds. The "masses" who, according to Weber's idealised understanding of domination, are supposed to assimilate the rules, do not take for granted what has been put in place for their own good. What prevails is a whole register of social performances aimed at circumventing a technology that threatens to render negotiation obsolete.

Over the years, Uganda has experienced a heavy burden of road traffic incidents. The country's related deaths are estimated at 28.9 per hundred thousand people compared to an average of 24.1 for Africa, and 18.0 worldwide (Balikuddembe et al. 2017). Accordingly, not less than three thousand victims were killed and fifteen thousand injured in road accidents every year between 2010 and 2013 (Uganda Police Force 2010, 2011, 2012, 2013). This is further reflected by the Ministry of Health statistics that place traffic injuries within the top ten causes of mortality in the country (Ministry of Health 2013). Drunk driving has been identified as a key factor accounting for over 40 percent of the road traffic incidents in Uganda (Mulondo 2013). Section 112 of the Uganda Traffic Control and Road Safety Act 1998 prohibits any person from driving a vehicle or operating any technical equipment which has consumed more than the legal limit of alcohol. In response, the police have turned to technology to aid their efforts and have rigorously introduced breathalysers to determine the blood alcohol concentration (BAC) of drunk driving suspects. This form of technicisation is applied during operations commonly referred to as Kawunyemu (breathe into it) which are usually conducted on the weekends roughly between 22h00 and 04h00. On the backstage, as Goffman (1988) would call it, only those officers who are to be deployed in a given operation are privy to the details, as one officer explained: "We have to out-smart the offenders who use all tricks to evade us (…) The key to a successful Kawunyemu is thorough preparation and control of information flow."

This particular form of technicisation simultaneously presents prospects and uncertainties. On the one hand, police officers appreciate that breathalysers help them to control the "luxurious crime" of drunk driving, yet on the other hand, those who are controlled perceive the technicisation as an encroachment on their lifeworld.[15] Some even suspect that this is another deceitful manipulation by the police to extort money. Or, as the journalist J.K. Abimanyi (2013) wrote, even "a drop" of alcohol could land one in jail. While there have been other methods of testing for drunk driving, such as asking suspects to walk on one foot, the breathalyser has made it easier for the police to provide a comprehensive legal basis for tackling the vice by removing their subjective judgement entirely.

The technicisation of Kawunyemu operations follows global strategies geared towards curtailing drunk driving. Their intensification in Uganda appears to have coincided with the World Health Organisation (WHO) that

---

15   For primary sources on this case see Kagoro (2022, chapter 5).

dedicated World Health Day 2004 to the issue of road safety (see WHO 2007)[16] and the 2011 proclamation of the United Nations Decade of Action for Road Safety 2011–2020. Drunk driving has been, and continues to be, framed as a social ill and many scholars have presented alarming accident statistics drawn from different parts of the world—all attributed to the role of alcohol (see Benson, Rasmussen, and Mast 1999; Eisenberg 2003; WHO 2007). It is difficult to know whether these operations have been successful in reducing drunk driving because the breathalysers used in Uganda do not automatically generate the necessary statistics. One-off statistical studies are not carried out, probably because they are expensive and not very reliable anyway.

In principle, the Uganda Police Force is able to prosecute drink-drivers much more effectively because the technology produces legally binding evidence. We argue that the police have embraced the breathalyser to become part of their professional lifeworld because it reinforces their official claim that they do nothing but enforce the law. Therefore, they find it easy to suspend the effort of making sense of the breathalyser, even without fully understanding how it actually works and why it should be trusted in the first place. But drivers who are told that the machine has found more alcohol in their breath than the law permits, are motivated to question what the officers take for granted. What they question is the objectivity of the measurement that, in their view, inexplicably comes out of a black box called a breathalyser. While the motive to question the reliability of the measurement seems logical and obvious, the arguments used for this purpose are worth paying closer attention to.

The police claim that there are many repeat offenders, some of whom have found ways to get around the roadblocks. Some police officers revealed that various strategies are common, such as setting up WhatsApp alert groups or even paying a *boda boda* (motorcycle taxi) to go ahead and warn the driver when they spot a roadblock. If none of these manoeuvres were available, drivers try to negotiate the issue. The argument about the reliability of the measurement usually begins after another attempt to avoid the fine. Knowing that they have been drinking and anticipating what might happen if they are stopped by the police, they quickly offer a bribe to the first officer who approaches them, as he or she is most likely to let them pass without testing after receiving the bribe. If this second type of manoeuvre fails, the test cannot be avoided. If the result shows a level of alcohol above the limit, a number of other tactics are used. Some drivers try to enact superiority and mock the officers; equally arrogant,

---

16   Proceedings marking this day were conducted in more than one hundred and thirty countries to raise awareness about road traffic injuries and stimulate new and improve existing road safety programmes (see WHO, 2007, xv).

others try to educate the officers to stick to their duties and respect the rights of citizens; a similar top-down approach is to argue that the machine must be faulty, sometimes adding that this is typical of the dysfunction of things in Uganda. At the other end of the spectrum, some drivers try to show equality and kinkeredness, either by using humour to change the officer's mood or by staging a drama. Some desperate drivers may try several such performances in a row. What all these theatrical and sometimes self-contradictory performances prove is that most of those who resort to these measures are somehow aware that there is more to it than the police technicising their corrupt ways of extorting money from citizens.

During our fieldwork, we came across a number of incidents where the stories drivers told after testing positive were based on the narrative construction of equality, trust, and drama. In one case, a driver in his late thirties urged the police to release him as soon as possible, saying that if they delayed, South Sudan would be doomed. The man claimed he had vital information that would help both Salva Kiir and Riek Machar find a peaceful solution to the conflict. Another man in his forties negotiated his way out by claiming, somewhat self-contradictorily, that his blood alcohol content (BAC) results were likely wrong because he had consumed a local brew that he claimed could not be detected by European-made machines. In another case, a woman in her early thirties who was driving what appeared to be an expensive car, and who was later identified as the wife of a wealthy businessman in the city, shouted instructions to the police: "(…) Don't waste my time because I am a breastfeeding mother, and I don't want to jeopardise my marriage! My husband does not know that I am out."

As mentioned above, the interesting thing about this and other similar stories is that they all carry the implicit message that the test result needs to be interpreted in the context of a particular, individual situation in order to make any sense at all. People in Uganda are used to constant negotiation and easily become sceptical when asked to blindly follow the rules as they are written. The breathalyser disrupts their normal expectation that there is always something to negotiate. And indeed, insisting that there is still room for negotiation is not as far-fetched as it might seem, given the lack of understanding of how the machine works. Simply accepting the results produced by a machine seems like unreasonable, blind submission. In addition, the pre-existing perception of a power asymmetry between police and citizens is reinforced when the machine further empowers the already more powerful side, and this constellation provokes resistance.

For our argument in this chapter, the important insight is that technicisation—the weaving of a technology into the taken-for-granted practices of a mundane lifeworld (see the Introduction to this volume)—can work differently in different lifeworlds of a society. For the Ugandan traffic police, we argue, the breathalyser has technicised their routine Kawunyemu operations. For many Ugandan drivers, certainly not all, the breathalyser has not become part of the unquestioned equipment of their lifeworld because it disrupts another and more important part of it, the understanding that nothing man-made can be non-negotiable.

3   What Have We Learned about Technicisation and Domination?

Numerification is generally understood as a form of technicisation that aims to codify and formalise social, economic, and political processes to make them more predictable, accountable, and transparent. In this chapter we have probed this understanding in the context of contemporary Uganda. At first glance, the various uses of numbers—in our case those generated by government budgets, police records, and breathalysers—appear to be familiar attempts geared towards what Max Weber called the rationalisation of the world. In our view, such an interpretation misreads Weber and consequently misinterprets the empirical reality it seeks to illustrate (see Boli and Thomas 1997; Bourdieu 2011, 108; Weber [1919] 1988, 594). This volume's engagement with STS and Blumenberg's argument about technicisation (see the Introduction to this volume) has encouraged us to take a step further in our year-long effort to develop an empirically grounded approach that avoids false teleological assumptions about the rationalisation of the world.

Drawing on the work of Theodore Porter and Emmanuel Didier, we argue that statistics and other forms of numerification do not simply represent the world, nor do they reduce it to simple categories that allow patterns to be seen. The production of numbers is always also a creative exercise, bringing to life something new that would not have been there without numerification (Porter 1995; Didier 2010). As the Introduction to this volume argues and we have shown empirically, when technologies—including forms of numerification—travel and are translated, they are inevitably re-inscribed with new meanings and sometimes even acquire new functions.

An important aspect of our study has been the question of how numerification becomes institutionalised as a particular form of domination—that is, how it becomes taken for granted, intertwined with the unquestioned routines of various entangled lifeworlds. We started from the undisputed classical insight that numerification is a form of domination, as Max Weber succinctly put it: "to rule by calculation" (Weber [1919] 1988, 594; our translation). Yet, technologies of governance are not only generated locally, but emerge from translations of globally circulating technologies (such as forms of numerification), significations, and even language games. Starting from these two basic assumptions, we have asked how ruling by calculation in Uganda works when it is nested in an ongoing global circulation of technologies of government. While the most prominent cases of such circulation are in the fields of health, environmental protection, and human rights, our interest was in policing and government budget control.

The introduction to this volume argued that the circulation of technology is based on translation, and that translation can be understood as the adaptation of a travelling technology to the networks of technologies, webs of institutions and beliefs (conceived of as lifeworlds) existing in the context into which the technology is translated. It has also emphasised that this is a bidirectional process in which the existing technology networks, institutions, and lifeworlds are transformed as the new technology is adapted to and internalised by them. Following this approach, we have shown in our three case studies, how, in the process of translating travelling technologies into new sites of practice through encounters and negotiations, what may have been hidden before becomes visible through the translation. It follows, that the link between technology and domination can be shaped in various ways. This implies that a certain level of standardisation is achieved across borders, making things legible for all those involved in the translation process—firstly for the trained personnel of the relevant government departments, and then for the citizens confronted with the rule. Yet, at the same time, and this is the main point of our chapter, the achieved level of standardisation and legibility does not automatically lead to the acceptance of the adapted technology as it is, nor to the internalisation of a new procedure in the everyday practices of governing or being governed. Our findings rather confirm and empirically prove what is argued in the Introduction to this volume (see also Rottenburg 2012, 2014).

In our first case study, we showed that budget support, as a travelling technology of internationalised government, has on the one hand easily become part of the established understanding that national government works through documents and measurements. The responsible Ugandan financial experts work with the budget as a form that they take completely for granted—they don't ask profound questions about its meaning, and have thus, in the terminology of this volume, suspended making sense of budgeting by using it as a naturally given part of the routines of their lifeworld. This state of affairs is largely due to the fact that budgeting as such has a long history in Uganda. On the other hand, while no one questions the necessity and rationality of national budgeting, budget support as a particular form of international agreements and dependencies remains controversial and was in fact abandoned after a few years. It has never been able to dominate the other imperatives that hover over the state officials themselves. The suspension of the meaning of budgeting, however, does not prevent Ugandan financial experts from questioning certain features and limits of the form of international budget support. It is therefore difficult to determine which parts of the budgeting process fade into the background as unquestioned realities, and which parts remain at the forefront of contestation.

Similarly, in our second case study, we were able to show that the case file was internally perceived and used in policing as a natural part of the working routine. We found the case file at the heart of the lifeworld of the police as a formal organisation. In other words, the life of the file became the life of the police. Unlike budgetary support and breathalysers, police records have come to be seen by Ugandan citizens as a self-evident and unquestioned part of their lifeworld. At the same time, and this is the most important part of our argument, we have been able to show that police records have taken on a function that no one intended. Instead of being used in legal proceedings following the initial police recording of an offence, both police officers and citizens have used them as evidence in informal conflict resolution and compensation settlements.

Our third case study, which looked at the introduction of breathalysers in the Ugandan traffic police, revealed another interesting aspect. Within the police force, they quickly became a welcome and taken-for-granted part of the daily routine of preventing drunk driving. At the same time, the use of breathalysers remained controversial among drivers and the general public, and therefore often became the focus of informal controversies and negotiations with police officers performing their duty. At stake was the very purpose of making alcohol testing non-negotiable by technicising the procedure in a context where anything controversial had to be negotiated. In the minds of most Ugandans, every case was seen as being fundamentally different and therefore the very idea of measuring cases with the same impartial and uncompromising technology seemed unfair and unreasonable.

In summary, our three case studies have shown that the question of technicisation—the extent to which a technology is woven into a lifeworld—needs to be differentiated. A first differentiation concerns the specification of the lifeworld into which a technology is or is not woven to become a habitual, unquestioned part of everyday practice. This was most evident in the third case study on the introduction of the breathalyser which showed a clear difference in habitualisation between traffic police and drivers. A second differentiation concerned the functionality of the technology once it had been woven into the fabric of two different lifeworlds. This was most evident in the second case study on the police file, as both the police, who are responsible for recording crimes brought to their attention, as well as the citizens of Uganda, normally take the idea of keeping police files for granted. In many criminal cases, from theft to road accidents and rape, most people in Uganda prefer compensation to long and often futile court cases, so few police files actually find their way into court, even though they are designed to do so. However, this situation does not diminish the presumed importance of police records outside the formal processes of law enforcement, as they have found a new important function in informal negotiations and settlements, with or without the involvement of the police. A third differentiation that our study brought to light concerned the need to specify the boundaries of the technology in question. Our first case study has drawn attention to the fact that a technology, in this case budgeting, can become fully entangled in a lifeworld, while at the same time not all of its uses and not all of its aspects need to be part of that entanglement. No one in the financial experts' lifeworld would question budgeting as the basis for running a government and governing a country, but the particular version of it, budget support, proved not to fit into the existing and unquestioned ways of doing things.

A classical modernist interpretation of these differentiations would probably point to practice, training, and the age of the technologies. While the files and the written documents have been part of practices and professional trainings since at least colonial times, both budget support and breathalysers are relatively recent introductions. These both require a certain rigidity in their application that is at odds with a social context that is much less formalised than contexts in which capitalism has long shaped minds and thoughts. In this sense, Marxist traditions and neoliberal modernisation theory seem to converge (Sohn-Rethel 1976; Archibugi and Michie 1995). In line with the other contributors to this volume and with the basic argument developed in our introduction, we argue that this interpretation is theoretically flawed and has been empirically refuted many times. The differences between the three trajectories of technicisation that we have observed in our case studies do not lend themselves to a ranking along the lines of more or less professional training and shorter or longer histories of practice and habitualisation. Rather than taking the level of technicisation as an indicator of modernity, we argue that its different levels are due to contradictory imperatives and logics that shape lifeworlds around the world in the process of globalisation.

Another important outcome of our contribution is that we have shown how combining elements of STS and political sociology can enhance the analytical leverage of the latter. Attention to travelling technologies and their translation into new contexts helps us to identify forms of domination in contexts where they have long been taken for granted and thus rendered invisible. In the case of practices of classification and numerification, this is all the more important and challenging as these are now taken for granted in both mainstream social science and government. Thus, if we assume that technicised domination is largely invisible, and if we further assume that the methods of the dominant social sciences are equally technicised, then the task of exposing opaque mechanisms of domination becomes particularly challenging.

Bibliography

Abimanyi, J.K. 2013. "Drink-Driving: Even a Drop of Alcohol Will Send You to Jail." *The Daily Monitor*, 2 March.

Akrich, Madeleine. 1992. "The De-Scription of Technical Objects." In *Shaping Technology*, edited by Wiebe Bijker and John Law, 205–24. Cambridge, MA: MIT Press.

Archibugi, Daniele, and Jonathan Michie. 1995. "The Globalisation of Technology: A New Taxonomy." *Cambridge Journal of Economics* 19: 121–40.

Balikuddembe Joseph Kimuli, Ali Ardalan, Davoud Khorasani-Zavareh, Amir Nejati, and Stephen Kasiima. 2017. "Road Traffic Incidents in Uganda: A Systematic Review of a Five-Year Trend." *Injury & Violence Research* 9, no. 1: 17–25.

Benson, Bruce L., David W. Rasmussen, and Brent D. Mast. 1999. "Deterring Drunk Driving Fatalities: An Economics of Crime Perspective." *International Review of Law and Economics* 19, no. 2: 205–25.

Biecker, Sarah. 2021. "People, Practices and Paper. An Ethnography of Everyday Policing in Uganda." Unpublished PhD dissertation, Department of Political Science, University of Bremen.

Biecker, Sarah, and Klaus Schlichte. 2015. "Between Governance and Domination: The Everyday Life of Uganda's Police Forces." In *The Politics of Governance: The State in Africa Reconsidered*, edited by Lucy Koechlin and Till Förster, 93–114. New York: Routledge.

Biecker, Sarah, and Klaus Schlichte. 2021. "A State of Numbers: Bureaucratic Technologies of Government and the Study of Internationalized Politics." In *The Political Anthropology of Internationalised Politics*, edited by Sarah Biecker and Klaus Schlichte, 177–98. Lanham, MD: Rowman & Littlefield.

Bijker, Wiebe E., Thomas Hughes, and Trevor J. Pinch, eds. 2012. *The Social Construction of Technological Systems: New Directions in the Sociology and History of Technology*. Cambridge, MA: MIT Press.

Boli, John, and George M. Thomas. 1997. "World Culture in the World Polity: A Century of International Non-Governmental Organization." *American Sociological Review* 62, no. 2: 172–88.

Bourdieu, Pierre. 2011. *Rede und Antwort*. Frankfurt am Main: Suhrkamp.

Callon, Michel. 2012. "Society in the Making: The Study of Technology as a Tool for Sociological Analysis." In *The Social Construction of Technological Systems: New Directions in the Sociology and History of Technology*, edited by Wiebe Bijker, Thomas P. Hughes, and Trevor Pinch, 77–97. Cambridge, MA: MIT Press.

Didier, Emmanuel. 2010. "Gabriel Tarde and Statistical Movement." In *The Social after Gabriel Tarde*, edited by Matei Candea, 163–76. London: Routledge.

Eisenberg, Daniel. 2003. "Evaluating the Effectiveness of Policies Related to Drunk Driving." *Journal of Policy Analysis and Management* 22, no. 2: 249–74

Foucarde, Marion. 2022. *Zählen, benennen, ordnen: Eine Soziologie des Unterscheidens*. Hamburg: Hamburger Edition.

Goffman, Erving. 1988. *Exploring the Interaction Order*. Cambridge: Polity Press.

Hull, Matthew S. 2003. "The File: Agency, Authority, and Autography in an Islamabad Bureaucracy." *Language & Communication* 23: 287–314.

Irish Aid. 2014. *Annual Report 2014*. www.irishaid.ie/media/irishaid/allwebsitemedia /20newsandpublications/publicationpdfsenglish/Irish-Aid-Annual-Report -2014-final.pdf.

Kagoro, Jude. 2022. *Inside an African Police Force: The Ugandan Police Examined*. Berlin: Springer.

Kjaer, Anne Mette, and Marianne S. Ulriksen. 2020. "The Political Economy of Resource Mobilization for Social Development in Uganda." In *The Politics of Domestic Resource Moblization for Social Development*, edited by Katja Hujo, 339–70. Cham: Palgrave for UNRISD.

Laiz, Álvaro Morcillo, and Klaus Schlichte. 2016. "Rationality and International Domination: Revisiting Max Weber." *International Political Sociology* 10, no. 2: 168–84.

Latour, Bruno. 1988. "Drawing Things Together." In *Representation in Scientific Practice*, edited by Michael Lynch and Steven Woolgar, 19–68. Cambridge, MA: MIT Press,

Latour, Bruno. 1999. *Pandora's Hope: Essays on the Reality of Science Studies*. Cambridge, MA: Harvard University Press.

Latour, Bruno. 2002: "Zirkulierende Referenz: Bodenstichproben aus dem Urwald am Amazonas." In *Die Hoffnung der Pandora. Untersuchungen zur Wirklichkeit der Wissenschaft*, 36–95. Frankfurt am Main: Suhrkamp.

Lie, Jon Harald Sande. 2015. *Developmentality: An Ethnography of the World Bank–Uganda Partnership*. New York and Oxford: Berghahn Books.

Ministry of Health. 2013. *Annual Health Sector Performance Report*. Kampala: Ministry of Health.

Mulondo, Lawrence. 2014. "40% of Road Fatalities Are Drunk Drivers." *New Vision*, December 17.

Parliament of the Republic of Uganda. 2013. *Report of the Committee on Defence and Internal Affairs on the Ministerial Policy Statements and Budget Estimates for the Fiscal Year 2013/2014*. Kampala: Parliament of the Republic of Uganda.

Popitz, Heinrich. 1992. *Phänomene der Macht*. Tübingen: Mohr Siebeck.

Porter, Theodor M. 1995. *Trust in Numbers: The Pursuit of Objectivity in Science and Public Life*. Princeton, NJ: Princeton University Press.

Riles, Annelise, ed. 2006. *Artefacts of Modern Knowledge*. Ann Arbor: The University of Michigan Press.

Rottenburg, Richard. 2012. "On Juridico-Political Foundations of Meta-Codes." In *The Globalization of Knowledge in History*, edited by Jürgen Renn, 483–500. Berlin: Max Planck Research Library for the History and Development of Knowledge.

Rottenburg, Richard. 2014. "Experimental Engagements and Metacodes." *Common Knowledge* 20, no. 3: 540–48.

Simson, Rebecca, and Bryn Welham. 2014. "Incredible Budgets: Budget Credibility in Theory and Practice." Working paper 400. London: Overseas Development Institute.

Sohn-Rethel, Alfred. 1976. "Das Geld, die bare Münze des Apriori." In *Beiträge zur Kritik des Geldes*, edited by Paul Mattick, Alfred Sohn-Rethel, and Hellmut G. Haasis, 35–117. Frankfurt am Main: Suhrkamp.

Thelen, Tatjana, Larissa Vetters, and Keebet von Benda-Beckmann. 2014. "Introduction to Stategraphy: Towards a Relational Anthropology of the State." *Social Analysis* 50, no. 3: 1–19.

Thompson, Gardner. 2003. *Governing Uganda: British Colonial Rule and Its Legacy*. Kampala: Fountain Publishers.

Uganda Police Force. 2010. *Annual Crime and Traffic/Road Safety Report 201*. Kampala: Uganda Police Force. www.upf.go.ug/download/publications(2)/Annual_Crime _Report_2010.pdf?x89335.

Uganda Police Force. 2011. *Annual Crime and Traffic/Road Safety Report 2011*. Kampala: Uganda Police Force. www.upf.go.ug/download/publications(2)/Annual_Crime _Report_2011.pdf.

Uganda Police Force. 2012. *Annual Crime and Traffic/Road Safety Report 2012*. Kampala: Uganda Police Force.

Uganda Police Force. 2013. *Annual Crime and Traffic/Road Safety Report 2013*. Kampala: Uganda Police Force. www.upf.go.ug/download/publications(2)/Annual_Crime _and_Traffic_Road_Safety_Report_2013(2).pdf.

Vismann, Cornelia. 2000. *Akten: Medientechnik und Recht*. Frankfurt am Main: Fischer.

Vismann, Cornelia. 2008. *Files: Law and Media Technology*. Stanford, CA: Stanford University Press.

Weber, Max. (1919) 1988. "Wissenschaft als Beruf." In *Gesammelte Aufsätze zur Wissenschaftslehre*, 582–613. Tübingen: Mohr.

Weber, Max. (1904) 2012. "On Some Categories of Interpretive Sociology." In *Max Weber: Collected Methodological Writings*, edited by Hans-Henrik Bruun and Sam Whimster, 273–300. New York: Routledge.

WHO (World Health Organization). 2007. *Drinking and Driving: A Road Safety Manual for Decision-Makers and Practitioners*. Geneva: Global Road Safety Partnership.

CHAPTER 6

Biometric Data Doubles and the Technicisation of Personhood in Ghana

Alena Thiel

1 Introduction

In recent years, governments across Africa, Latin America, and South Asia have intensified their investments in digital national identification systems (Gelb and Clark 2013; Breckenridge 2010, 2014). Ghana, the case study of this chapter, consolidated this trajectory of datafying her population with the introduction of the biometric identity card in 2003 and its eventual nation-wide roll-out from 2017. Datafication, here, entails the selection and abstraction of social life for the purpose of putting it "in quantified form so that it can be tabulated and analysed" in an automated, large-scale, and value-generating fashion (Couldry and Mejias 2019, 2). The datafication of personhood, then, relies on the production of "data doubles" which are pieces of information about a body that come to stand in as its "doubles" in computational environments. Doubles circulate in the form of data strings that encode specific combinations of categorical variables supposed to capture the person and to tie these to a unique body (Haggerty and Ericson 2000; Bouk 2018).

Official efforts at datafying the Ghanaian population through biometrics have since come to encompass the national identification system, health and social security insurance registers, passports, drivers licensing, and voter registration. Increasingly, these biometric registers are mutually integrated while also connecting across registrations of financial and telecommunication transactions, thus expanding the potential for datafied personhood. In 2021, the National Health Insurance and the Social Security and Pension Fund both discontinued their respective ID numbers. Indexicality was instead achieved through adoption of the biometrically authenticated Personal Identification Number or PIN of the National Identification Authority. More recently, mobile money accounts and the registration of SIM cards have also been tied to the national ID register. Linked to the biometric ID, Ghanaians are further datafied in space through the novel GPS-based digital address system which ties the national ID to unique data points in space (Thiel 2023).

This chapter explores the ways in which these recent and still emerging efforts to expand the datafication of the Ghanaian population through biometrics impact expressions of personhood in people's lived, day-to-day, interactions. Beyond doubt, the institutional settings producing datafied bodies—within Ghana's national identification and statistical bodies—present distinct "lifeworlds" of technical experts that, in turn, are composed of particular participatory routines and taken for granted assumptions, that deserve attention in and off themselves. The focus of this chapter will however be on the effects of technicised personhood on the social life of ordinary Ghanaians. The chapter's goal, in other words, is to analyse the far-reaching, yet often invisible effects that ensue when technologies are *translated* into what Michael Jackson (2014, 2017) has termed the lifeworld, or the human perception of the world and one's being-in-it. To do so, the chapter retraces the ways in which new, automated forms of identification affect everyday routines of performing one's status in relation to others (both people and institutions) to the point of these routines receding into the background of the encounter. The chapter hence does not offer a normative analysis of whether technological innovation improves experiences of citizenship or claims to particular rights—including their subversive counter-expressions (Cakici, Singh, and Thiel 2024). Instead, it provides a detailed view into how these technologies integrate and interlace with multiple lifeworlds upon circulation between contexts of production and reception.

The idea that a unique, administratively registered identity is fundamental for inclusion in citizenship goes back to the UN Convention of the Right of the Child (1989) which defines a unique identity—established through birth registration—as a universal right and fundamental function for inclusion in a wide range of other rights. The Millennium Development Goals (MDGS) further underscored this idea, while the Sustainable Development Goals (SDGS) for the first-time defined identity registration as a development goal in and for itself (SDG 16.9). Development practitioners have linked biometric technologies to empowerment and citizenship, for example as a response to statelessness (Manby 2017). Yet the idea that biometric technology can provide individuals the possibility to prove their unique identity and ultimately claim inclusion in a political community cannot be uncritically adopted. Cautioned by the historical example of the Nazi census (Aly and Roth 2004), scholars have pointed to the potential threats of centralised databases for minority populations (Labbé 2009; Seltzer and Anderson 2001), while biometrics have been shown to perpetuate discrimination (Ajana 2012; Rao 2018) and even denationalisation (Salem 2018). In light of this, Marielle Debos (2021) reminds us that biometric technologies can never be politically neutral, nor can they in and for themselves fix social or political conflict. For this reason, further analyses of identification systems' sociotechnical implications remain crucial.

Critical studies of the datafication of populations (von Oertzen 2017) have experienced a strong revival following earlier canonical interest by French and American sociologists of quantification (Desrosières 2000; Porter 1995). Studies focusing on data politics (Isin and Ruppert 2019), material-semiotic glances on data infrastructuring (Pelizza 2016, 2021), data practices (Cakici, Ruppert, and Scheel 2020), as well as the performativity of data in its temporal (Sepkoski 2018) and visual-aesthetic forms (Ratner and Ruppert 2019), have made major advances in conceptualisations of how datafication is tied to the making of populations and of informational personhood alike (Koopman 2019; Ruppert 2008). The debate about population data infrastructuring maintains a clear regional bias however, focusing mainly on North Atlantic cases. Szreter and Breckenridge (2012, 17) problematise the failure to represent the "diverse history of community registration" and the "profound bias in the documented comparative historical record, which provides us mainly with a history of registration which appears to be strongly tied to those most powerful, persisting state-like forms of government which generated and archived most efficiently the records of their processes of registration." As African census and civil registration systems are significantly shaped by their colonial histories (Ittman, Cordell, and Maddox 2010; Thiel 2022), they require a critical examination of the continuities of hegemonic data histories in contemporary relations of power and subjectivity.

In line with the focus of this volume, I draw here on Hans Blumenberg's (2015) understanding of the immersion of technology in the lifeworld to trace how the formalisation of identification processes in Ghanaian state bureaucracy has reduced the identification encounter to what Blumenberg refers to as "mere method." Identification technologies achieve this by virtue of replacing the associative and dialogical establishment of unique identities in interpersonal encounters with the conventionalised measurement of particular bodily dimensions. In their conventionalised forms, biometric identification systems reiterate the grid of categories that establish equivalences between individuals, sorting them into social groups, ranking them at times, all while concealing this process in this very automatism. By addressing how these data practices intersect with everyday lifeworlds, the conceptual shift pursued here can shed a new light on the contemporary concern of the unevenness of post-colonial data politics (Isin and Ruppert 2019; Madianou 2019; Couldry and Mejias 2018).

The chapter proceeds by first elaborating on the idea of the technicisation of personhood. It then illustrates empirically how technological innovation alters the configuration of variables establishing formally documented personhood in Ghana.[1] The chapter finally revisits the question of identification through the theoretical lenses of technological change and emerging *technicisation*.

2 Technicising Personhood

The central claim of this chapter is that technological changes impact the production of personhood, to the effect that the process of identification retreats into the background of consciousness. Edward Higgs (2009, 353) illustrates the effect of technological change on historical forms of establishing formal identities. According to Higgs, "techniques and technologies of identification have always been implicated in the formation and perception of identity in society." Technological transformations in the field of identification, Higgs (2009) shows, are therefore not new additions, somehow externally layered onto society's natural state, but have historically shifted from body markers (such as tattoos or scarification) to linking bodies with proxies that exist outside of them and can be processed independently. I show here that although the interpersonal relations at the heart of establishing formal identities have long been subsumed under these methods, current biometric technologies render the production of personhood even more opaque to the thus identified person.

Technicisation, according to Blumenberg, describes the age-old process by which meaning (*Sinnbildung*) morphs into method (and by extension, technology), and which then no longer requires interpretative acts of sense-making but reproduces itself in the form of mere function (Blumenberg 2015, 185). This process, Blumenberg argues, has become part of the modern consciousness and outlook onto the world (2015, 163, 169). As a result, method and function are perceived as unquestioned realities (*wahres Sein*) rather than products of this historical process (2015, 165, 185). Technicisation in this sense cannot be reduced to the notion of the antithetical relationship between technology and nature. Rather for Blumenberg, who critically draws on Husserl's concept of the lifeworld—understood by him as the taken-for-granted framework from which

---

1   Empirical data for this chapter was collected during four consecutive fieldwork periods in Ghana from January to February 2016, July to August 2017, April to May 2018, and January to March 2020. In addition to long-term media observation on the Ghanaian e-ID system, the data collected consisted predominantly of expert interviews conducted in the various government agencies involved in the production of Ghana's National Identification ecosystem. Interviews were carried out with legal, financial, and technical experts of the project, staff of the previous registration exercises, activists, and civil society organisations, as well as citizens involved in the previous registration exercise (Greater Accra and Central Region).

experiences and cognitions derive their sense or meaning—technology cannot oppose the natural lifeworld since it is itself already a part of it (2015, 184).

Following Blumenberg further, with the immersion of technology in the lifeworld, technology becomes an unquestioned given and loses its appearance of contingency (2015, 178). Unlike Husserl, however, Blumenberg does not consider this moment as being constitutive of a loss of meaning, but rather suggests that it entails a deliberate eschewal. Technicisation, in Blumenberg's argument, relieves social actors from the need to understand and repeatedly reproduce complex applications, and instead allows "unreflected repetition" irrespective of the user's insights into the operation (2015, 194). As users merely trigger effects that are already vested in the apparatus—rather than consciously generating them—the contingency of the construction is concealed, and the apparatus is legitimised in its mere presence. As a result of this emancipation of technology from signification, human action becomes homogeneous and unspecific, reduced to triggering formalised functions (Blumenberg 2015, 188, 190).

The chapter proceeds to lay out how formalisation of personhood through biometric technologies represents a process of technicisation. Formalisation occurs, for example, in relation to Ghanaian pluralistic naming conventions that are not only socially coded but also highly situationally contingent. In the moment of biometric identification, names are fixed in a universally formatted, single official record. In contrast to the interpersonal "dance" of establishing relations of hierarchy, respect and care through cultural naming practices, these processes of constructing formal personhood recede into the background of the lifeworld in the very moment the identification encounter is automated and reduced to the recording of a fingerprint, iris, or personal identification number. As James Scott, John Tehranian, and Jeremy Mathias (2002, 4) noted, modern state practices of identifying the person seek to overlay relational forms of establishing personhood with their "standardised administrative techniques" in the form of birth registration and the assignment of a unique identity expressed in a name coupled with basic biographic statistics (birth date and place, parental relationship). Following Scott et al.'s argument, codifications of naming by states arrive at "taming the chance" of illegibility, making "[q]uestions of inheritance, paternity, and household affiliation far more transparent" (2002, 9,10) as they technicise the social universe in which identification is situationally and relationally defined.

Biometric identification technologies "associate identities to individuals by using measurable personal features instead of something owned or known by the individual" (Mordini and Massari 2008, 488). Through fast-paced technological advancements, biometric technologies can rely on an expanding set of indicators, including behavioural biometrics that measure the body's interaction with its environment, and often combine several of these indicators in a multi-modal approach (Mordini and Massari 2008; see also Pelizza 2021). In this reduction of personhood to the (biometric) data double, I argue, everyday routines of performing one's status in relation to others (including institutions) become immersed in automated routines, receding thus into the background of the encounter. To remain with the example of naming, where a Ghanaian name bears important information about a person's status, origin, and belonging,—all of which participate in the encounter between persons and often institutions—in the biometric authentication of identity, names are completely void of meaning. Biometrics, in other words, silence the myriad of practices that go into the dialogical establishment of identity and its expression in personhood, including the possibility for critical engagement. In short, as biometric technologies are translated into new contexts such as Ghana, they automate previously foregrounded encounters to the effect of rendering these interactions increasingly opaque. Fundamental to this dynamic is the integration of the technology into the lifeworld, where it's operation and underlying assumptions are both invisible and in the background of the machine's operation while at the same time exerting influence over the lifeworld through the world-making qualities of classifying and hence making up personhood (in the eyes of selves and others).

3 The Technicisation of Personhood in Ghana

Classic anthropological thought has linked the foundation of personhood in Ghana to cosmological considerations, extending the realm of personhood into spiritual and inanimate dimensions (Fortes 1987). More recent scholars of identity in Ghana have especially focused on ethnic identification, among others in relation to different forms of conflict over resources such as land and political influence (Lentz 2006; Valsecchi 2001; Tonah 2009, see also Schildkrout 1978), but also in relation to culture (Schramm 2000) or even sports (Darby 2013; Agyekum 2019). In addition to ethnic group identities, scholars describe Ghanaian identities in economic terms—for example in descriptions of the rising middle class, or the analysis of elite formation (Osei 2015). So far, however, surprisingly little attention has been paid to how technological changes affect Ghanaian notions of identity.

Recent contestations about the survey instrument of the 2021 decennial Population and Housing Census (PHC) revealed the continued relevance of the identity question. Ghana's most recent census saw intensive contestations around categories of ethnicity in several parts of the country. Tensions around issues of representation and inclusion of certain ethnic groups boiled up in the Volta region around the alleged failure to list subgroups of the Ewe ethnicity (since this had been done for all other major ethnic groups). Public commentary further collated these allegations with partisan alliances, since the Volta region has historically been associated with the opposition party NDC. Crucially, the 2021 PHC triggered public debate and commentary about who counted as Ghanaian, alleging that certain groups listed in the survey instrument represented foreigners who should be excluded from the count. The common motive of dismissing residents of the Eastern borderlands as Togolese was reproduced in these narratives, in addition to falsely dismissing other groups, such as Fulani and Hausa populations, as immigrants and hence non-Ghanaians (Thiel 2021b). As Carola Lentz (2008) illustrated, membership in the Ghanaian colonial and postcolonial state has historically been subject to contextual negotiations of boundaries, interests, and associations. The contestations around the 2021 census categories dramatically rearticulated partisan affiliations, ethnic identities, and historically contested national boundaries (Nugent 1995). The fixing of categorical variables in the technology of the census, in other words, rearticulated intensive contestation around the question of identity (see also Hacking 1995).

While census questionnaires still relatively easily allow respondents to grasp the categories mobilised to classify them based on their answers, the recent proliferation of biometric identification technologies has rendered these categories much less transparent, so much so, that they retreat into the background of the encounter between citizens and authorities. As a result of their invisibility, these technologies—which themselves participate closely in world-making, along with dominant conceptualisations of self and other—thus become engrained in the lifeworld of Ghanaians and hence their modus of conceiving and being in the world. The following parts of this chapter will be concerned with how the automation of identification, that is the process of linking a body to a unique set of variables such as name and birth dates, strips away this moment of rendering visible the categories which "make up" populations and personhood in Ghana.

3.1   The Biometric Ghanacard: Reshaping Identification through Data Innovations

Biometric technologies differ from existing practices of identifying personhood in their opaqueness, or in the words of Blumenberg, in their reduction to mere method. Some Ghanaian communities traditionally inscribe identifiers in the body through scarification (Bowles 2009). Another form of embodied identification mainly concerns traders in fresh produce who often have their names and hometown, sometimes even a phone number, tattooed on their bodies for enabling repatriation in case of accidents during their frequent sourcing trips to remote farms.[2] However, very little attention has been paid to more recent practices of identifying personhood in Ghana in the form of officially standardised, quantifiable categories of personhood.

Keith Breckenridge (2010) notes how in 1972, Ghanaian military leader General Ignatius Acheampong introduced the first national identification documents in Ghanaian history. Whereas this initiative remained limited to the country's ethnically overlapping border regions, from 1987 onwards, the Provisional National Defence Council of Flight Lieutenant Jerry John Rawlings sought to resuscitate this idea and expand its reach nationwide. A number of working committees formulated technical proposals which were eventually picked up in the National Economic Dialogue and, in 2001, a call for tender proposals was published to develop a national identity registration infrastructure (Breckenridge 2010, 645). A comprehensive civil identity registration project was then set in motion with the establishment of the National Identification Authority Secretariat in 2003 (Breckenridge 2010, 646; Odartey-Wellington 2014; Akrofi-Larbi 2015; Owusu-Oware et al. 2017; Thiel 2015, 2020). With the recent consolidation of the Ghanaian ID system and its penetration into ever more realms of life, we begin to observe effects on the constitution of personhood in interpersonal and state-citizen interactions.

Biometric technologies identify a person not on the basis of documentary evidence but on matching (by way of a probabilistic relation) physiological measurements—like for instance a fingerprint or iris scan—and its earlier registration in a database. This reduction of the identification encounter, development economists have argued, can provide an alternative to the painstaking task of building the national infrastructure of civil registration systems where these systems are historically underfunded and lacking in coverage (Szreter and Breckenridge 2012). As Breckenridge (2021) has shown, the problem with this approach is that it does not automatically link the body to the social person. In practice, therefore, national identification authorities such as Ghana's NIA face difficulties in establishing beyond doubt the identity of registrants in the moment of biometric mass registrations (Thiel 2020). In view of this, in November 2017, a group of experts—among them representatives of Ghana Statistical Service, Births and Deaths Registry, and National Population Council—came together to call for the technical and institutional integration

---

2   Unlike the argument put forth by Higgs (2009), this bodily inscription does not imply any
    association with delinquency.

of biometric identity registration with records of civil registration, in particular, the national Births and Deaths register. Through the infrastructural link of a life-long and centralised personal identification number (PIN), the policy planners noted that identities could then be carried forth through various life events, marked across a range of administrative registers relating to health, education, or the labour market, as a means for recording the interaction of citizens and the state in various areas of life (Thiel 2020).[3]

In addition to fixing personhood through biometric measurement and tracing personhood through the PIN across such administrative government registers, Ghanaian government officials sought to locate formal personhood in space by further linking identification with a new digital addressing system. While "houses do not have fingerprints" and address data is based on GPS-coordinates rather than biometrics, the compulsory recording of an address at the moment of registration for the new Ghanacard has advanced the construction of a nationwide address register. Specifically, the 2017 revision of the National Identity Register Act (which provides the legal basis of the biometric register) expanded the required addressing information for registration in the national identification system to include not only street or post box addresses, but also the digital address code assigned through the GhanaPost GPS app. Consecutively, other population registers were joined through biometrics and the PIN. In the fall of 2021, the National Health Insurance Card, Social Security and Pensions Insurance Card, and the electronic tax ID were all integrated under the National Identification Authority's PIN, thus marking the phasing out of these duplicate ID numbers and cards. Simultaneously, a compulsory link between the SIM card and the NIA register was established by the Ministry of Communication, with further plans to extend an additional link to a national equipment register of mobile devices. And with increasing penetration of the PIN, further infrastructures and interfaces were subsequently layered onto the ID system. One example was the recent announcement of an app-based interface allowing employers to check the Ghana Police criminal investigation records through their candidates' PIN. This expansion of datafication of

---

3   The underlying data logic of these initial plans was to retrospectively strengthen civil registration and vital statistics along with the introduction of the new mode of identifying personhood. Yet turf wars between the institutions involved in these high-level consultations, under the direct supervision of the Vice Presidency, resulted in the side-lining of the birth and death registration. First, this occurred because several alternatives to the birth certificate were accepted as feeder documents—hence undermining the birth register as a basic register. Second, the PIN was assigned at the moment of biometric identity registration through the NIA rather than by the Births and Deaths Registry.

personhood has important ramifications for the ways in which individuals and their life-courses become legible as datafied entities.

In the late 2000s, Ghanaians met the introduction of the *ezwich* debit card, the world's first biometric payment system (Breckenridge 2010), with outright suspicion, falsely blaming the system's opaqueness in transactions for losses of funds (Dzokoto, Asante, and Aggrey 2017). Similarly, years of attempting to introduce biometric IDS left Ghanaians tired of repeated registration exercises and the increasing politicisation of the system (Thiel 2020). But with the Covid-19 pandemic hitting Ghana, digital identification for remote transactions found a decisive catalyst. Fear of infection as well as imposed restrictions in mobility and assembly expanded the business opportunities for newcomer, app-based delivery services in urban areas—such as Glovo or Jumia—which carry meals to offices and deliver online shopping and postal deliveries. These services fell on fertile ground in the digitally literate urban middleclass, where banking, addressing, and identification systems had previously penetrated and hence facilitated such remote transactions. Mobile money services further expanded interactions at a distance, triggering an upsurge in agent kiosks along the country's roadsides, and more sophisticated multi-platform dealers offering specialised financial services at strategic sites in urban neighbourhoods. Along with this expansion of mobile financial services, efforts to regulate the market intensified too. For this reason, all mobile money services require card-based identification from April 2021, and since the Electronic Transfer Levy (Amendment) Act 2022 (ACT 1089) was passed, are taxed at a rate of 1 percent.

Formally identifying personhood has had further effects on state-citizen interactions. A recent empirical example is the biometric terminal of the Social Security and National Insurance Trust in Accra (SSNIT), where pension and social insurance holders can manage their accounts and claims in an automated encounter process authenticated via a biometric interface (Figure 6.1). Provided environmental conditions such as the climate's influence on fingerprints' readability allow the functioning of the machine,[4] the SSNIT biometric terminal completely removes Ghanaian bureaucrats from the transaction process.

In other words, such automated processes provide clients with the possibility to update records, such as the legally mandated annual submission of a life certificate required for pension payments, through their SSNIT smartcard and PIN, rather than facing a customer agent in person. Identification technologies are thus said to enhance the formalisation of encounters, reducing spaces

---

4   Factors such as the level of humidity and racial engineering biases have a significant bearing on false response rates (Rao 2018; Pelizza 2021).

FIGURE 6.1   Advertisement for the SSNIT biometric terminals
              SOCIAL SECURITY AND NATIONAL INSURANCE TRUST N.D.

for discretion,[5] corruption, but also potentially discriminatory practices. This argument is particularly relevant in view of the large body of literature on African bureaucracies (Bierschenk and Olivier de Sardan 2014) which reminds us of the various ways in which African state writing practices (see Biecker et al., this volume) are personalised and shaped by complex translations and boundary work (Beek and Bierschenk 2020). In view of such gains in accountability, remote interactions with the Ghanaian state have been deliberately encouraged in the ongoing, presidentially acclaimed Ghanaian Data Revolution. As a central part to this agenda, the Ghanaian eGov online portal encourages Ghanaian citizens to apply for permits, certificates, school placing, or filing taxes with one and the same PIN (Ghana.gov, n.d.).

---

5   In practice, biometric technology does not completely rule out incidents of interpretation and the application of discretions. Especially in the moment of the initial registration of the body, what is considered evidence of personhood remains contested.

Cutting across these applications, biometrics promise a one-size-fits-all approach to transparency, accountability, and improved planning. While the widespread adoption of biometric solutions is only gradually consolidating across the various sections of the Ghanaian population data system, the technology has effectively rendered invisible the categories mobilised to produce official personhood. As the process of automation of identification through biometrics takes centre-stage, associated processes of classifying and sorting have receded into the background of the technicised lifeworld. In all this, identification technologies are however not mere tools, but exert important disciplining effects—such as on (hetero)normative conceptions of personhood (Thiel 2021a). The following section digs deeper into the social effects of rendering the categories of personhood invisible in the automation of identification.

3.2 *The Political Consequences of Technicised Identification*

Biometric identification processes conceal the relational nature of identification by reducing it to mere method. In this section, I discuss the political consequences of the technicisation of the identification encounter. Innovation processes often constitute largely conservative projects (Suchman and Bishop 2000). Database innovations in particular carry forth path dependencies in that they build on infrastructural arrangements of material devices, software, personnel, and forms of organisation (Edwards et al. 2007). This observation is particularly relevant for understanding the persistence of "ghost variables" making up personhood in Ghana. Ghost variables, Rebecca Karkazis and Katrina Jordan-Young (2020, 764–66) argue, relate to histories of abusive systems which haunt the everyday by "reaching deeper than the conscious perception". In line with this, African census historians have shown in great detail how categories of ethnicity, religion, labour and health not only contributed to establishing colonial power but also continued to shape postcolonial nation-building projects in line with these hegemonic projects (Isin and Ruppert 2019, Ittman, Cordell and Maddox 2010). Engin Isin and Evelyn Ruppert (2019, 214) note how colonial means of knowledge production (maps, censuses, identification) have a continued influence as "technologies of colonial government of counting, categorising, and ordering were inherited, reshaped, and reused by postcolonial governments." In particular, the authors note, ethnic and racial categories relate to the imagination of the nation in ways instituted by the former imperial powers (Isin and Ruppert 2019, 208).

In Ghana categories of ethnicity remain part of state knowledge production and the production of personhood. The contestations around representation of ethnicity in the recent census as laid out above illustrate the longue durée of such colonial categories. Another example occurs in the moment of birth registration where parents' ethnicity is recorded for statistical purposes. As my encounters with state officials have shown, in such instances, fathers from certain ethnic groups often face discriminatory attitudes towards their expected contribution to the livelihood of their children. However, these reproductions of essentialising normative frames can no longer be questioned when the categories of the person registered recede into the background of the automated verification of biometric patterns and numeric IDS, which hide the very categories used to establish personhood. This is because, in the words of Karkazis and Jordan-Young (2020, 733), "[r]eckoning with ghosts requires being in relation with them," a process that arguably is rendered impossible with increasing technicisation.

Nonetheless, the concept of the lifeworld can offer a valuable entry point to the question of how we can rearticulate critical responses to the datafication of personhood in the current moment of postcolonial data politics. Drawing on anthropological readings of the lifeworld, in particular Jackson's (2014; 2017) take on this concept, the role of affection in the construction of the human perception of the world and one's being-in-it has a certain relevance. Following Jackson, conceptualising the retreat of technology into the lifeworld is complicated by the understanding that the lifeworld itself is constructed in the oscillation "between moments of complete absorption in an immediate situation and moments of detachment" and reflection (Jackson 2014, 27). The case study of Ghana underscores Jackson's point that technicisation represents a highly contingent process, which can include actors actively stepping back for the observation of their place in the world.

The example of naming practices again is useful in making apparent these oscillations between absorption and emotional detachment. In a recent upstir in Ghana's Parliament, the proposal of the Death and Birth Registry to eradicate names like "Maame" or "Auntie" from the birth register was met with highly emotional reactions about preserving cultural and highly affective practices of naming children after honoured members of the community. Moments like this reveal what is lost or goes unquestioned in the process of technicisation. To fully appreciate the extent of this, a more extended discussion about the nature of personhood in moments of technological change is necessary.

4   Personhood and Identity through the Lens of Technology

Questions around constructions of personhood, and by extension about selfhood and identity, have received intensive academic interest for many years.

Anthropological debates in particular have a long history of conceptualising personhood, pointing to the vast variations in which human culture assigns status, rights, and moral obligations through relational mechanisms such as ancestry or affiliation. As a category of the mind or form of consciousness, personhood, Marcel Mauss' classic work established, is closely tied to our notions of morality. It is fundamental for our collective life in defining our legal, political, economic, and religious positionalities vis-à-vis others (Mauss [1938] 1985). This consciousness of the self, Georg Herbert Mead expands, is rooted in the ongoing social process of interaction within which "mind and selves arise" (as summarised in the introduction by Charles Morris; see Morris 1972, xv). In Mead's pragmatist view, personhood arises experientially through a social process in which (self)consciously communicating individuals internalise emerging, generalisable symbols within a community of meaning (Morris 1972, XXVI, 138).

Building on these essentially relational concepts of personhood, Engin Isin and Patricia Wood (1999, 19) argue that the process of identification can be conceptualised as the intersubjective recognition of a person's unique status in the form of a dialogical establishment of attributes constituting "an index of individual position and disposition [...], a classification that places and positions an individual within a social space." As a practice of association (Isin and Wood 1999), identification therefore represents a fundamentally political concept, especially where demands for recognition and inclusion in citizenship are concerned (Isin and Wood 1999, 13; Krause and Schramm 2012).

4.1   Data Subjects, Data Doubles, and Human Kinds

As a result of its political weight, identification finds its expression in complex bureaucratic processes. Edward Higgs (2009) traces the historical development of material technologies that facilitate the exchange of credentials through which individuals claim particular statuses in social and political institutions. According to Higgs, this establishment of recognised personal attributes has shifted from interpersonal forms of truth making, such as oaths, to material technologies, including seals, signatures, and—only more recently— identity documents and biometrics (Higgs 2009, 346). The adoption of increasingly automated forms of identification through biometrics decouples the body from documentary, biographic forms of evidence about personhood. In its stead, socially empty, arithmetic forms of identification establish uniqueness and associate with the social identity only in a later, non-essential step of linking biometric and documentary records (Breckenridge 2021).[6]

---

6   As Colin Koopman reminds us, the datafied manifestation of "informational selfhood" does not point to an "essential self beneath all of our data" (Koopman 2019: 19) but undergoes

The technicised codification of personhood, things and actions represents an "investment in forms" that is essential for social coordination, evaluation, and ranking (Boltanski and Thévenot 1991). In particular, modern practices of targeting individuals for state intervention, Ian Hacking (1985) has shown canonically, are modelled on the idea that societies must seek to classify human behaviour into ever more refined categories, or "human kinds", in the pursuit of intervening in and modifying their behaviour (Hacking 1995, 351–52, 360). Following Hacking (1995, 352), "human kinds [are] kinds about which we would like to have systematic, general, and accurate knowledge; classifications that could be used to formulate general truth about people; generalisations sufficiently strong that they seem like laws about people, their actions, or their sentiments (...) precise enough to predict what individuals will do, or how they will respond to attempts to help them." For Hacking, human kinds "are part of what we take knowledge to be. They are also our system of government, our way of organising ourselves" (Hacking 1995, 364). The power of human kinds, Hacking notes crucially, rests in them being taken for granted and receding into the background of people's lifeworld.

In order to produce such human kinds that go unquestioned, quantitative conventions of generalisation need to create equivalence between individual variations, thus enabling their classification (Ruppert 2008). As Evelyn Ruppert (2008, para. 2.1) notes, quantification of the population works by eradicating individual variation through creating equivalence based on "classifying and identifying difference and resemblance to numerous categories (male, female, married, single etc.) in relation to a pre-formatted classification grid (sex, marital status, racial origins etc.)" (Ruppert 2008, para. 2.2, 2.3). Individuals are thereby made "comparable and commensurable in all their difference" (Ruppert 2008, para. 2.4). For Ruppert, the process of dividing and differentiating the population into such categories is fundamental to the production of the population, as these very categories "connect the individual to the population, hold people together and collectively, once assembled, constitute the population" (Ruppert 2008, para. 2.3). This coordination (and ranking) of categories within the larger entity of the population eventually facilitates the formal positioning of personhood between individuals and vis-à-vis institutions. However, categories do not always befit the thus classified populations and they only recede into the background of consciousness when they are perceived as

---

complex sequences of material-semiotic "translations" linking the person to her data double (Pelizza 2021).

natural and undisputed (Ruppert 2008). The introduction of automated identification, I have argued, renders this type of dispute less likely.

5   Conclusion

The functioning and impact of emerging biometric systems and their immersion in the lifeworld remains fragile, potentially disrupting the technicisation
of personhood in its course. Technicisation has been defined by Blumenberg
(2015) as the process by which functions are formalised into mere method.
Standardisation and rationalisation of the process allow endless empty rep-
lication without requiring users' insight into the operations underlying the
application. As this process involves the retreat of technology's functioning
into the background of users' consciousness, technology's immersion into the
lifeworld is contingent on its smooth operation.

The complex, relational, and fluid interpersonal encounters of identification becomes formalised in biometric technologies. In this process, emphasis
moves away from documenting the unique combination of categorical variables that establish formal personhood and instead foregrounds measurement-
based procedures of determining whether the person presenting herself is
identical with the body that was initially registered. In this process, the categories associated with formal personhood become entirely invisible to the user.

Yet, in their invisibility, these categories of personhood continue to exert
their disciplining force. As new modes of computation potentially shift the
ways in which categorical variables are produced and recombined, it becomes
crucial to shed critical light on how the possibilities for personhood are
affected. This is particularly relevant as ethnic and gendered classifications
carry within them histories of colonial domination. The technicisation of personhood, I have tried to show, further black boxes these already largely invisible genealogies, thus haunting contemporary Ghanaian society in ways that
are increasingly difficult to resist. At the same time, the expansion of digital
identification technologies has been presented in this chapter as a manifestation of postcolonial data politics, and the capture of African population
data systems by a handful of North-Atlantic technology-providers. Adopting
a dynamic understanding of the lifeworld, in which actors oscillate between
immersion and critical reflection can lead us to appreciate how the inherent
frictions in these systems may refocus our attention on the largely invisible
processes of "making up" personhood and thus to open up new modalities of
political engagement.

Bibliography

Agyekum, Humphrey Asamoah. 2019. "The Best of the Best: The Politicisation of Sports under Ghana's Supreme Military Council." In *Sports in African History, Politics and Identity Formation*, edited by M. Gennaro and S. Aderinto, 73–88. London: Routledge.

Ajana, Btihaj. 2012. "Biometric Citizenship." *Citizenship Studies* 16, no. 7: 851–70.

Akrofi-Larbi, Richmond. 2015. "Challenges of National Identification in Ghana." *Information and Knowledge Management* 5, no. 4: 41–45.

Aly, Götz, and Karl Heinz Roth. 2004. *The Nazi Census: Identification and Control in the Third Reich*. Philadelphia, PA: Temple University Press.

Beek, Jan, and Thomas Bierschenk. 2020. "Bureaucrats as Para-ethnologists: The Use of Culture in State Practices." *Sociologus* 70, no. 1: 1–17.

Bierschenk, Thomas, and Jean-Pierre Olivier de Sardan. 2014. *States at Work: Dynamics of African Bureaucracies*. Leiden: Brill.

Blumenberg, Hans. 2015. "Lebenswelt und Technisierung unter Aspekten der Phänomenologie." In *Schriften zur Technik*, 163–202. Berlin: Suhrkamp.

Boltanski, Luc, and Laurent Thévenot. 1991. *On Justification: Economies of Worth*. Princeton, NJ: Princeton University Press.

Bouk, Dan. 2018. "The National Data Center and the Rise of the Data Double." *Historical Studies in the Natural Sciences* 48, no. 5: 627–36.

Bowles, Laurian R. 2009. "Imaging Migrant Women and the Embodied Market: Accra, Ghana." In *Expressions of the Body: Representations in African Text and Image*, edited by Charlotte Baker, 313–36. Bern: Peter Lang Verlag.

Breckenridge, Keith. 2010. "The World's First Biometric Money: Ghana's e-Zwich and the Contemporary Influence of South African Biometrics." *Africa: The Journal of the International African Institute* 80, no. 4: 642–62.

Breckenridge, Keith. 2014. *Biometric State: The Global Politics of Identification and Surveillance in South Africa, 1850 to the Present*. Cambridge: Cambridge University Press.

Breckenridge, Keith. 2021. "Documentary Government and Mathematical Identification: On the Theoretical Significance of African Biometric Government." In *Identification and Citizenship in Africa: Biometrics, the Documentary State and Bureaucratic Writings of the Self*, edited by Séverine Awenengo Dalberto and Richard Banégas, 49–64. Abingdon, Oxon: Routledge.

Cakici, Baki, Evelyn Ruppert, and Stephan Scheel. 2020. "Peopling Europe through Data Practices: Introduction to the Special Issue." *Science, Technology, & Human Values* 45, no. 2: 199–211.

Cakici, Baki, Ranjit Singh, and Alena Thiel. 2024. "The Politics of Seamlessness: A Rights Claims Perspective on Digital Identification Technologies." In *Digital*

States in Practice, edited by Jessamy Perriam and Katrine Meldgaard Kjær. Berlin: De Gruyter.

Couldry, Nick, and Ulises Mejias. 2019. "Data Colonialism: Rethinking Big Data's Relation to the Contemporary Subject." *Television and New Media* 20, no. 4: 336–49.

Darby, Paul. 2013. "'Let Us Rally Around The Flag': Football, Nation-Building, and Pan-Africanism in Kwame Nkrumah's Ghana." *The Journal of African History* 54, no. 2: 221–46.

Debos, Marielle. 2021. "Biometrics and the Disciplining of Democracy: Technology, Electoral Politics, and Liberal Interventionism in Chad." *Democratization* 28, no. 8 : 1406–22.

Desrosières, Alain. 2000. *La Politique des Grands Nombres. Histoire de la Raison Statistique*. Paris: La Découverte.

Dzokoto, Vivian, Rebecca Asante, and John K. Aggrey. 2016. "Money That Isn't: A Qualitative Examination of the Adoption of the 1 Pesewa Coin and Biometric Payment Cards in Ghana." *Ghana Studies* 19: 3–34.

Edwards, Paul N., Steven J. Jackson, Geoffrey C. Bowker, and Cory P. Knobel. 2007. "Understanding Infrastructure: Dynamics, Tensions, Design." Report of a Workshop on "History & Theory of Infrastructure: Lessons for New Scientific Cyberinfrastructures". Accessed March 27, 2022. www.ics.uci.edu/~gbowker/cyberinfrastructure .pdf.

Fortes, Meyer. 1987. *Religion, Morality and the Person: Essays on Tallensi Religion*. Cambridge: Cambridge University Press.

Gelb, Alan and Julia Clark. 2013 "Identification for Development: The Biometrics Revolution." Working Paper 315. Washington, DC: Center for Global Development. http:// dx.doi.org/10.2139/ssrn.2226594.

Ghana.Gov. n.d. "Ghana's Digital Services and Payments Platform." Accessed March 27, 2022. www.ghana.gov.gh.

Hacking, Ian. 1985. "Making Up People." In *Reconstructing Individualism: Autonomy, Individuality, and the Self in Western Thought*, edited by Thomas C. Heller, Morton Sosna, and David E. Wellbery, 222–36. Stanford, CA: Stanford University Press.

Hacking, Ian. 1995. "The Looping Effects of Human Kinds." In *Causal Cognition: A Multi-Disciplinary Debate*, edited by Dan Sperber, David Premack, and Ann Premack, 351–83. Oxford: Clarendon Press.

Haggarty, Kevin, and Richard Ericson. 2003. "The Surveillant Assemblage." *The British Journal of Sociology* 51, no. 4: 605–22.

Higgs, Edward. 2009. "Change and Continuity in the Techniques and Technologies of Identification over the Second Christian Millennium." *Identity in the Information Society* 2, no. 3: 345–54.

Isin, Engin, and Evelyn Ruppert. 2019. "Data's Empire: Postcolonial Data Politics." In *Data Politics: Worlds, Subjects, Rights*, edited by Didier Bigo, Engin Isin, and Evelyn Ruppert, 208–27. London: Routledge.

Isin, Engin, and Patricia Wood, 1999. *Citizenship and Identity*. London: Sage.

Ittman, Karl, Dennis Cordell, and Gregory Maddox. 2010. *The Demographics of Empire*. Athens: Ohio University Press.

Jackson, Michael. 2014. "Ajàlá's Heads: Reflections on Anthropology and Philosophy in a West African Setting." In *The Ground Between: Anthropologists Engage Philosophy*, edited by Veena Das, Michael Jackson, Arthur Kleinman, and Bhrigupati Singh, 27–49. Durham, NC, and London: Duke University Press.

Jackson, Michael. 2017. *How Lifeworlds Work: Emotionality, Sociality and the Ambiguity of Being*. Chicago and London: Chicago University Press.

Karkazis, Katrina, and Rebecca Jordan-Young. 2020. "Sensing Race as a Ghost Variable in Science, Technology, and Medicine." *Science, Technology, & Human Values* 45, no. 5: 763–78.

Koopman, Colin. 2019. *How We Became Our Data: A Genealogy of the Informational Person*. Chicago: University of Chicago Press.

Krause, Kristine, and Katharina Schramm. 2012. "Thinking Through Political Subjectivity." *African Diaspora* 4, no. 2: 115–34.

Labbé, Morgane. 2009. "Statistique Ethnique, Légitimité Politique et Changement de Régime." *Critique Internationale* 45: 9–18.

Lentz, Carola. 2006. *Ethnicity and the Making of History in Northern Ghana*. Edinburgh: Edinburgh University Press.

Lentz, Carola. 2008. "'This Is Ghanaian Territory!': Land Conflicts on a West African Border." *American Ethnologist* 30, no. 2: 273–89.

Madianou, Mirca 2019. "Technocolonialism: Digital Innovation and Data Practices in the Humanitarian." *Social Media and Society*, July–September 2019: 1–13.

Manby, Bronwen. 2017. *The World's Stateless Children*. Oisterwijk: Wolf Legal Publishers. https://files.institutesi.org/worldsstateless17.pdf.

Mauss, Marcel. (1938) 1985. "A Category of the Human Mind: The Notion of Person; The Notion of Self." In *The Category of the Person: Anthropology, Philosophy, History*, edited by Michael Carrithers, Steven Collins, and Steven Lukes, 1–25. Cambridge: Cambridge University Press.

Mordini, Emilio and Sonia Massari. 2008. "Body, Biometrics and Identity." *Bioethics* 22, no. 9: 488–98.

Morris, Charles. 1972. "Introduction". In *Mind, Self, and Society. From the Standpoint of a Social Behaviorist*, by George Herbert Mead, ix–xxxv. Chicago and London: University of Chicago Press.

Nugent, Paul. 1995. *Big Men, Small Boys and Politics in Ghana: Power, Ideology and the Burden of History, 1982–1994*. London: Pinter.

Odartey-Wellington, Felix. 2014. "A Technological Invasion of Privacy: The Need for Appropriate Responses to the New Surveillance Society in Ghana." *Ghana Center for Democratic Development Briefing Papers* 13, no. 4. https://cddgh.org/vol-13-no -4-technological-invasion-of-privacy-the-need-for-appropriate-responses-to-the -new-surveillance-society-in-ghana-by-felix-odartey-wellington.

Osei, Anja. 2015. "Elites and Democracy in Ghana: A Social Network Approach." *African Affairs* 114, no. 457: 529–54.

Owusu-Oware, Emannuel, John Effah, and Richard Boateng. 2017. "Institutional Enablers and Constraints of National Biometric Identification Implementation in Developing Countries: The Case of Ghana." In *Twenty-Third Americas Conference on Information Systems, Boston. AMCIS 2017: Proceedings*: 13. https://aisel.aisnet.org /amcis2017/ICTs/Presentations/13.

Pelizza, Annalisa. 2016. "Developing the Vectorial Glance: Infrastructural Inversion for the New Agenda on Government Information Systems." *Science, Technology, & Human Values* 41, no. 2: 298–321.

Pelizza, Annalisa. 2021. "Identification as Translation: The Art of Choosing the Right Spokespersons at the Securitised Border." *Social Studies of Science* 51, no. 4: 487–511.

Porter, Ted. 1995. *Trust in Numbers: The Pursuit of Objectivity in Science and Public Life.* Princeton, NJ: Princeton University Press.

Rao, Ursula. 2018. "Biometric Bodies, or How to Make Electronic Fingerprinting Work in India." *Body and Society* 24, no. 3: 68–94.

Ratner, Helene, and Evelyn Ruppert. 2019. "Producing and Projecting Data: Aesthetic Practices of Government Data Portals." *Big Data & Society* 6, no. 2: 1–16.

Ruppert, Evelyn. 2008. "'I Is; Therefore I Am': The Census as Practice of Double Identification." *Sociological Research Online* 13, no. 4.

Salem, Zekeria Ahmed. 2018. "Touche Pas à ma Nationalité: Enrôlement Biométrique et Controverses sur L'identification en Mauritanie." *Politique Africaine* 152, no. 4: 77–99.

Schildkrout, Enid. 1978. *People of the Zongo. The Transformation of Ethnic Identities in Ghana.* Cambridge: Cambridge University Press.

Schramm, Katharina. 2000. "The Politics of Dance: Changing Representations of the Nation in Ghana." *Africa Spectrum* 35, no. 3: 339–58.

Scott, James, John Tehranian, and Jeremy Mathias. 2002. "The Production of Legal Identities Proper to States: The Case of the Permanent Family Surname." *Comparative Studies in Society and History* 44, no. 1: 4–44.

Seltzer, William, and Margo Anderson. 2001. "The Dark Side of Numbers: The Role of Population Data Systems in Human Rights Abuses." *Social Research* 68, no. 2: 481–513.

Sepkoski, David. 2018. "Data in Time: Statistics, Natural History, and the Visualization of Temporal Data." *Historical Studies in the Natural Sciences* 48, no. 5: 581–93.

Suchman, Lucy, and Libby Bishop. 2000. "Problematizing 'Innovation' as a Critical Project." *Technology Analysis & Strategic Management* 12, no. 3: 327–33.

Szreter, Simon, and Keith Breckenridge. 2012. "Recognition and Registration: The Infrastructure of Personhood in World History." *Proceedings of the British Academy* 182: 1–36.

Thiel, Alena. 2015. "Entangled Temporalities: Ghana's National Biometric Identity Registration Project." *Anthropology Today* 33, no. 1: 3–5.

Thiel, Alena. 2020. "Biometric Identification Technologies and the Ghanaian 'Data Revolution'." *The Journal of Modern African Studies* 58, no. 1: 1–22.

Thiel, Alena. 2021a. "Biometric Payment and Gendered Kinds in Ghana." *Tapuya: Latin American Science, Technology and Society* 4: 1–17.

Thiel, Alena. 2021b. "Ghana Upgraded Its Census to Make it More Inclusive: But Old Tensions Still Surfaced." *The Conversation,* July 22, 2021. https://theconversation.com/ghana-upgraded-its-census-to-make-it-more-inclusive-but-old-tensions-still-surfaced-164654.

Thiel, Alena. 2022. "Une histoire politique du recensement au Ghana aux 20e et 21e siècles." *Statistique et Société* 10, no. 1: 59–78. https://publications-sfds.fr/index.php/stat_soc/article/view/845.

Thiel, Alena. 2023. "Digital Addressing and the Construction of the Ghanaian Nation-Space." *Digital Geography and Society* 4: article 100054. https://doi.org/10.1016/j.diggeo.2023.100054.

Tonah, Steve. 2009. "Democratization and the Resurgence of Ethnic Politics in Ghana, 1992–2006." In *Ethnicity, Belonging and Biography: Ethnographical and Biographical Perspectives*, edited by Gabriele Rosenthal and Artur Bogner, 63–82. Berlin: LIT.

Valsecchi, Pierluigi. 2001. "The 'True Nzema': A Layered Identity." *Africa: Journal of the International African Institute* 71, no. 3: 391–425.

Von Oertzen, Christine. 2017. "The Historicity of Data: Concepts, Tools, and Practices during the Nineteenth Century." *Zeitschrift für Geschichte der Wissenschaften, Technik und Medizin* 25: 407–34.

CHAPTER 7

Project Time, Lifetime, and Extra Time: Technicisation of Mass HIV Treatment Programmes and the Acceleration of Pharmacy in Uganda

Sung-Joon Park

1   Introduction

In this chapter, I use Hans Blumenberg's notion of technicisation to explore the expansion of mass HIV treatment programmes in Uganda.[1] In my analysis of the technical in mass HIV treatment, I will discuss the acceleration of time in what has been called the "scale-up of access to treatment" in order to provide a critique of global health. The acceleration of time discussed in this chapter differs from the anthropological critique of the power of technologies, such as indicators (Adams 2015; Rottenburg et al. 2015), and more recent scholarly concerns about the pervasive imperatives of speed in global health, and more generally in science (Fortun 1998; Adams, Burke, and Whitmarsh 2014; Stengers 2017). I suggest that an ethnographic exploration of the expansion of mass HIV treatment—enabled by life-saving pharmaceuticals and a range of technologies to manage their logistics—is salient for a critique of the acceleration of time in global health.

Drawing on Michel Foucault's notion of biopower, anthropologists have shown that technologies might have ensured access for poor nations to life-saving pharmaceuticals, yet by simplifying treatment, they reduce subjects to bodies, cells, and pills (Prince 2014; Sangaramoorthy and Benton 2012). We cannot fully understand these forms of subjectification by hinging our critique of technologies on the simplified representations of the self and the world they produce. As Wendy Espeland (2015, 61) aptly notes, indicators are: "Technologies of simplification, strategies that make complex processes visible and easy to grasp, and make comparisons—across people, organisations, or time—easy. This simplification is both why we value indicators so much and why we often feel they misrepresent us."

---

1   My discussion specifically draws on the essays *Technicisation and Lifeworld* published in 1963 under the German title "Lebenswelt und Technisierung unter Aspekten der Phänomenologie" (1963) and Blumenberg's 1986 book *Lebenszeit und Weltzeit* (Blumenberg 1986).

In my view, Espeland's observation elucidates Blumenberg's phenomenological critique of what he calls "technicisation" in his essay *"Lebenswelt und
Technisierung unter Aspekten der Phänomenologie"* (Blumenberg [1963] 2015).
Blumenberg uses the term technicisation to refer to the relentless expansion
of technical objects. However, his critique of technicisation departs from the
classical phenomenological thought that modern technologies deprive subjects of an unabridged experience of the world at hand. Instead of juxtapos-
ing the growing technicisation of the world against the lifeworld, Blumenberg
proposes that technologies, like indicators, are necessary for us to simplify our
relation to the world even if we end up feeling the relationship between self
and world is misrepresented. In the case of indicators, the simplification of this
relationship, as Espeland (2015) argues, involves an erasure of stories through
which we make sense of our world and ourselves as subjects. But this erasure
relieves us of the need to constantly reflect on the consequences of each and
every action, as Blumenberg interjects.[2]

He reminds us to be wary of the phenomenological critique of technicisation, which presumes that technology alienates us from the lived experiences
of the world, which he polemically calls a "backward-looking grievance"[3]
about the "loss"[4] of an unabridged experience of the world (Blumenberg [1963]
2015, 1986). According to Blumenberg, this form of critique is characteristic for
the phenomenological tradition inspired by Edmund Husserl's notion of phenomenology, which is constituted by a pre-theoretical lifeworld from which
scientific and technological abstractions are derived. This phenomenological
sentiment of a loss of meaning ignores that history, including the expansion of
technical objects, is irreversible (Blumenberg [1963] 2015).

Certainly, the point that history is irreversible has been made by other
thinkers too. However, Blumenberg elaborates this subtle point more radically
towards a critique of acceleration of time amplified by processes of technicisation (Blumenberg 1986). More importantly, in view of contemporary scholarly
debates about acceleration, he argues early on that acceleration is not necessarily experienced as an increase in speed, but as a shortage of time. This shortage of time, in contrast to the reflection on the value of time by historian E.P.
Thompson (1967), is produced by a growing gap between lifetime and world
time that is characteristic of the modern age (Blumenberg 1986).

This critique of acceleration, as I seek to show in this chapter, is useful for
the assessment of global health. I will first discuss how Blumenberg's notion

---

2 Blumenberg calls this erasure or waiver of sense-making practices in German *Sinnentlastung*.
3 *Sinnverlustklage* in German.
4 *Sinnverlust* in German.

of technicisation provides crucial clarifications for a critical anthropology
of speed and acceleration in global health. Secondly, drawing on my ethnographic research on the logistics of antiretrovirals in Uganda, I explore how
global health makes the acceleration of access to HIV treatment possible by
translating *access* into a technical problem—namely supplying large amounts
of medication to a population without losing time. This translation is achieved
through indicators and donor-funded projects that produce HIV treatment
on a larger scale. If one takes a cue from Blumenberg's distinction between
lifetime and world time, the flipside of accelerating access to treatment is the
yawning gap between what I call *project time*—the time limited by the dura-
tion of project—and *lifetime*, which denotes how time is spent as a mode of
becoming (Blumenberg 1986). World time, represented abstractly in hours or
years, insinuates that the temporal horizon for achieving global health through
donor-funded projects is infinite. Project time reminds us that project goals—
those to be achieved within a given time frame—are not just idealistic imagi-
nations of a better world but profoundly ignore the lived experience of time
as the most basic dimension of being with others. Thirdly, I explore how acceleration not only leads to narratives through which people make sense of the
world and themselves being erased, but also—more radically—the possibility of gaining insight being undermined. *Insight*, a term used by Blumenberg
without further explication, requires *extra time*, as my Ugandan interlocutor
tellingly put it.[5] This extra time aptly captures how global health progress is
not to be mistaken for scientific and social transformations based on insight.
In this chapter, insight means that knowing *what* global health or pharmacy is
arises out of learning *how* global health or pharmacy is done. My discussion
of extra time focuses on the pharmaceutical profession, specifically mass HIV
treatment, and illuminates how the shortage of time reflects a general transformation of pharmacy that, according to pharmacists, has little to do with
pharmacy as a profession.

2   Acceleration and the Problem of Insight

Blumenberg (1986) provides a number of systematic and polemic arguments
against the juxtaposition of technical world and lifeworld (see also Introduction). What interests me most is Blumenberg's own critical phenomenology
of technicised lifeworlds and how this phenomenology may contribute to the
anthropology of global health. He does not provide an optimistic philosophy

---

5   Fieldnotes, October 31, 2010.

of technologies. He is rather a philosopher of history who asks how history can account for the immense acceleration of time, which technicisation of everyday life relentlessly amplifies. In doing so, he relinquishes the critique of instrumental rationalities for an analysis of acceleration as a distinctive epoch in human history. Blumenberg may not have fully elaborated his critical inquiry of acceleration beyond the elongated discussion of how history can be included in the phenomenological analysis of technicisation—an aspect that Husserl failed to recognise, as he argues. Nevertheless, in view of contemporary scholarly debates on acceleration (for instance Rosa 2013) and recent anthropological discussions of speed and acceleration (Adams, Burke, and Whitmarsh 2014; Pigg 2013), Blumenberg's early engagement with acceleration as a "pathological" dimension of technicisation might be quite useful to recuperate the critique of global health (Blumenberg [1963] 2015, 202; see also Fassin 2013).

The first critical reflections on acceleration are found towards the end of his essay *Technisication and Lifeworld*. He asserts that development, as used by conventional development theory, can be seen as a geographical expansion of technicisation. Blumenberg's view on this form of technicisation differs from his preceding discussion of Husserl's notion of lifeworld. He notes that development is presented as a leap[6] forward in the sense of making social, economic, or political progress. However, without an intrinsic want for this progress, development means nothing other than catching up temporally with the so called industrialised world under the tutelage of a rather abstract idea of modernisation (Blumenberg [1963] 2015, 201–2).

Blumenberg introduces the notion of *social acceleration* quite early on and goes beyond the commonplace understanding that technologies increase the pace of life—to attend to the historical conditions turning time into a scarce resource (Blumenberg 1986; see also Rosa 2013). In his book *Lifetime and World Time,* he elaborates on his analysis of the relationship between technicisation and acceleration by tracing the growing incongruence between lifetime and world time throughout history.[7] This gap underlies our experience that lived time is in short supply in view of a seemingly infinite temporal horizon of future possibilities promised by modernity. Blumenberg's conclusion is that

---

6   In various other contexts, Blumenberg uses the German word *Sprung,* here translated as "leap". Technical devices enable us to step out of the realm of lifeworldly taken-for-granted truths that otherwise might hold us back from gaining other insights than those already given.

7   *Zeitschere* in German.

scholarly controversies about technicisation are misguided by grappling with the problem of agency in a technicised world. Instead, as he suggests, acceleration shows that "efficiency" and "insight"[8] constitute the more fundamental antinomy evinced by technicisation (Blumenberg [1963] 2015, 201). That is, acceleration achieved through greater technical efficiency makes an embodied understanding of technologies obsolete.

According to these remarks, insight stresses that *knowing-that* cannot be separated from *knowing-how* (Blumenberg [1963] 2015, 168). In the case of this chapter, the concept of insight highlights that knowing what global health or pharmacy is about cannot be separated from how it is done. It provides an important departure to move beyond the critique of speed in global health toward a critical analysis of the ways the speed of doing global health erases the time for producing insights.

The sociologist Hartmut Rosa, providing the most comprehensive social theory of acceleration, infers from Blumenberg's observation of a growing incongruence of lifetime and world time, that acceleration cannot be countered by the reduction of speed, as various popular voices are demanding (Rosa 2016). Recent anthropological studies have devoted great attention to the notion of acceleration for opening up new sites for critical studies of global health (e.g. Adams, Burke, and Whitmarsh 2014, 30). The body of scholarly literature on acceleration is diverse and heterogeneous, maintaining the elusiveness of a common definition of the term (Fortun 1998). Within this body of literature, Blumenberg's contribution is to distinguish between speed and acceleration, or better, between technical acceleration and the social acceleration of time. This distinction is crucial to go beyond acceleration achieved through technical innovation—some of them being vital, such as, in this case, the expansion of large-scale HIV treatment programmes—and to move towards a critical study of the incongruence between lifetime and world time. In the field of global health intervention, this incongruence is grasped as a gap between lifetime and project time. This gap is increasingly experienced as a lack of time, created by the demand to use time efficiently, as measured by indicators, within a project to achieve global health goals—as opposed to the time needed to learn how to do things; this time is experienced as time spent with others.

In the rest of the chapter, I show how an inquiry into acceleration can be fruitful for an analysis of mass HIV treatment programmes that contributes theoretically and empirically to a critical ethnography of global health. To this

---

8 In the original text, efficiency and insight are termed *Leistung* and *Einsicht*.

end, I will begin with an ethnographic account of indicators measuring treatment numbers and discuss what technicisation and acceleration means. In doing so, I want to elaborate an understanding of the technicised lifeworld prompted by the infrastructure of mass HIV treatment by tracing the gaps between lifetime and project time. I will specifically focus on pharmacists who play a pivotal role in managing the infrastructure of large-scale HIV treatment programmes as logistical experts. My analysis of logistics in the pharmaceutical field suggests that acceleration—defined as a shortage of time for generating insight—requires *extra time*. As my interlocutors aptly pointed out, this extra time is in short supply in the field of global health. Extra time is time spent beyond the technical terms of references defined by projects—the "Terms of References"—whose primary objective is to increase efficiency and improve accountability. Consequently, we may ask to what extent global health is not only lacking money, resources, or knowledge, but crucially, extra time for producing insight.

3   Indicators and Progress in Global Health

In the early years of antiretroviral therapy, the demand for medication for all who needed it gave rise to broadening access being equated with speed. Programmes were expressly described as "accelerating access to treatment". The well–known humanitarian relief organisation Médecins Sans Frontières issued reports entitled "Speed Saves Lives," reminding one of the urgencies of the HIV epidemic in an attempt to speed up the roll–out of antiretrovirals, especially in African countries (Redfield 2013, 69).

The Millennium Development Goals (MDGS) provide an internationally agreed indicator or measure of progress in terms of providing access to antiretroviral therapy. Under these goals, which were established in 2000 as a framework for development, all United Nations (UN) member states endorsed the goal to provide universal access to HIV treatment by 2015 (WHO 2003). Universal access does not mean providing everyone with antiretroviral medicines. Rather, it is an indicator that measures and tracks the actual number of patients on treatment in relation to those who are eligible for treatment, vaguely defined as those "in need".

Instead of using this vague definition, international treatment guidelines recommend the use of CD4 counts to simplify treatment decisions. According to this recommendation, a patient is eligible for treatment when their CD4 count falls below a certain threshold. In the first years of antiretroviral therapy, in countries such as Uganda, a CD4 count of 150 was the threshold at which treatment was started. This threshold determines the number of people eligible for treatment—65 thousand out of approximately 1.2 million people were living with HIV in Uganda in 2003—which constitutes the total number of people to be provided with HIV treatment under the goal of universal access to treatment (Okero 2003). Over time, the threshold was increased to a CD4 count of 250 and then 350 because of medical reasons and, more importantly, because it became economically and technically feasible to start treatment long before CD4 counts dropped below these thresholds.

In Uganda, like in various other African countries, initiatives to accelerate access to treatment led to significant reductions in—and even the total absence of—costs for antiretroviral therapies. When antiretrovirals became free, access to the medication increased rapidly. In Uganda, the number of patients on antiretrovirals increased from an estimated 4,000 in 2004 to an estimated 270,000 in 2009 (Whyte et al. 2006). The move to make medication free for users followed a dramatic reduction in global prices for antiretrovirals. The number of donor organisations and the amount of donor aid also increased. The Ugandan budget of United States President's Emergency Plan for AIDS Relief (PEPFAR) and the Global Fund to Fight HIV/AIDS, Malaria, and Tuberculosis was double the national health budget.

As a consequence, the absolute number of patients to be treated under the universal access goal increased too, namely to more than 700,000 people in Uganda, and this goal was still to be reached by 2015. In 2015, this was only half of the estimated total number of 1.2 million people living with HIV in the country. Globally, the number of people living with HIV has been more than 32 million in these years (UNAIDS 2013) and most of them live in sub-Saharan Africa.

Some anthropologists have become wary about the persistent demand to do more and to do it faster. These anthropologists criticise the idea of acceleration for being a manifestation of the "neoliberal ethics of speed and efficiency" (Pigg 2013, 2). Vincanne Adams, Nancy Burke, and Ian Whitmarsh promote the idea of slow science to raise a critical awareness about doing science with care, making deliberate choices, and—most importantly—taking time to conduct research (Adams, Burke, and Whitmarsh 2014). Stacey Leigh Pigg (2013) reminds us that ethnographic research unfolds in long time spans necessary for noticing the invisible changes of the world, which is increasingly shrinking under a neoliberal ethics of speed and efficiency.

Thinking with Blumenberg, this anthropological critique of acceleration is not necessarily questioning the need to speed up the provision of access to treatment. Rather, it directs our attention to the profound contradiction between efficiency and insight that is unravelled by acceleration, as Blumenberg put it. By contrast, a narrow critique of speed obscures that technicisation and the acceleration of time prompted by technicisation are irreversible. The critique of acceleration needs to consider that global health produces a shortage of time by framing access to antiretroviral therapy as a technical problem.

In the following section, I give an ethnographic account of the accelerated expansion of large-scale HIV treatment in Uganda and argue that progress towards universal access has not been too slow but rather too fast. I pay specific attention to the "projectification" of care and treatment in Uganda (Whyte et al. 2013). In doing so, I focus on pharmacists who play a pivotal role in broadening access to treatment as logistics experts. My ethnographic account of the technicised lifeworld focuses on the notion of extra time, proffered by the participants, to explore how acceleration is achieved at the expense of insight. Insight, which requires extra time, suggests that care is more than the provision of treatment and requires time that is in short supply in global health.

3.1 *More Patients*

When I first met Immanuel Mutabaazi, the senior pharmacist at The AIDS Support Organisation (TASO) in 2010, his organisation had just raised the overall target from 33,000 to 37,500 patients on treatment. Funded by international donors, TASO and other NGOS played a major role in reaching the target of universal access to treatment in Uganda. When I asked Immanuel how the increase had come about and specifically if it was the result of increased donor funding, he responded:

> No, not at all. In fact, our funding for drugs has remained stable since we started. We were only realising savings through price reductions. We started with five hundred patients in 2005. We only ask our funders if we can increase the number of patients when we make savings. Then we set our new targets.[9]

The main way in which TASO accelerated access to treatment was through reducing prices of the most widely used standard first–line antiretroviral regimes. The calculation of financial savings and the adjustment of treatment slots—the technical term for the number of patients a programme can take on—requires no more than a Microsoft Excel table containing patient numbers, drug prices, and so on (Figure 7.1).

Immanuel described this practice of increasing the number of treatment slots through price reductions as a "culture of savings" that TASO and other

---

9 Interview IM, Kampala, Uganda, January 14, 2010.

TASO UGANDA CDC 2010 ARVS BUDGET

Clients ARVS

Total number of Clients:	32,990
Percent on first-line treatment:	98%
Percent on second-line treatment+Pead	2%
Percent adults:	100%
Percent children:	0%

1st line treatment (adults)	Percent	Number of Clients
Total number of Clients	32,330	
	100%	32,330
	100%	

2nd line treatment (adults)	Percent	Number of Clients
Total number of Clients	660	
	100%	660
	100%	

Clients Enrolment within 12 months

						Generics			$8.64 Unit Cost 33,552		$73,87 Unit Cost 132,966
Months	%	#days	Clients	Client Days	Client Months	1st Line		Cost	2nd Line		Cost
1		360	32,990	11,876,400	395,880	387,962		1,084,742.870	7,918		87,730.967
2		330						1,084,742.870			87,730.967
3		300						1,084,742.870			87,730.967
4		270						1,084,742.870			87,730.967
5		240						1,084,742.870			87,730.967
6		210						1,084,742.870			87,730.967
7		180						1,084,742.870			87,730.967
8		150						1,084,742.870			87,730.967
19		120						1,084,742.870			87,730.967
10		90						1,084,742.870			87,730.967
11		60				-		1,084,742.870	-		87,730.967
12		30						1,084,742.870			87,730.967
	0%		32,990	11,876,400	395,880	387,962		13,016,914.445	7,918		1,052,771.602

Total ARVS-Old Centres $ 14,069,686.046

8,276.286

Additional slots   11,571

GRAND TOTAL   8,276.286

FIGURE 7.1   Excel table used by TASO to calculate savings and recalculate patient numbers

WITH COURTESY FROM ISAAC MUTABAAZI, TASO

NGOs, such as the Joint Clinical Research Centre and the Makerere Joint AIDS Program,[10] had developed over time.

As Immanuel emphasised, savings were the result of working closely together with the local pharmaceutical supplier Medical Access, which employed a range of experts in pharmaceutical supply chain management with a background in pharmacy or medicine. They strategically used procurement practices to lower the cost of antiretrovirals within the existing legal and political framework set by intellectual property rights, international and national drug regulations, and donor organisations' interests (T'Hoen et al. 2011; Peterson 2014).

As a result, from 2001 onwards, the prices for antiretrovirals declined dramatically, from $10,000 per patient per year to less than $250 per patient per year in 2009. Andrew Sowedi, the owner of Medical Access, explained that the largest price reductions had come from switching from expensive brand-name antiretroviral regimens to cheaper generics. After all branded antiretrovirals had been replaced by generic antiretrovirals, the procurement strategy mainly aimed at comparing drug prices in neighbouring countries, and more widely, by using global drug price indicators such as those documented in Médecins Sans Frontières' "Untangling the Web of Price Reductions". Buying in bulk also meant lower prices, Andrew said.[11]

Treatment slots is the technical term used to indicate the number of patients a treatment programme is able to accommodate. In the beginning, international treatment guidelines simplified which patients qualified for treatment by recommending the use of CD4 counts as a prerequisite. From a logistical point of view, however, the number of patients who could be enrolled for treatment was determined by the amount of antiretrovirals a programme could reliably procure with a finite budget provided by global health donor organisations.

Increasing access to treatment essentially amounted to negotiating price reductions and thereby increasing the number of slots. This was particularly effective in the first few years of free antiretrovirals. In 2009, however, no further price reductions were available for standard first–line therapies—they had already reached their lowest prices. In this year, Medical Access was able to bring the prices down to $20 per pack, which amounted to a saving of $0.50 per pack. It might not seem much but for Immanuel this was enough to add 4,500 new treatment slots to TASO's overall target. However, these additional slots had to be distributed across all eleven TASO treatment centres in the country,

10   These are the largest providers of antiretroviral therapy in the country.
11   Fieldnotes, October 28, 2010.

which meant each centre could in effect only treat fewer than five hundred additional patients. According to Immanuel this "was nothing'" compared to the real demand at all TASO sites.

From 2004, PEPFAR funding had not increased substantially, as Immanuel pointed out. The prominent Ugandan HIV activist and physician Peter Mugyenyi feared that the US government was aiming to "flatline" its funding for antiretroviral therapy (Mugyenyi 2009). The national health budget was hardly able to maintain all patients on treatment after the Global Fund to Fight AIDS, Tuberculosis, and Malaria had suspended its grant in 2005 over unresolved allegations of corruption (Taylor and Harper 2014).

In 2010, in view of a staggering demand for treatment and antiretrovirals frequently being out of stock, the Ministry of Health convened an emergency meeting to present its most recent findings on the funding gap for HIV treatment (Figure 7.2).

After prices could not be reduced any further, it was only possible to increase the number of treatment slots if the allocated budget was increased. Indicators are attractive technologies that produce simplified numerical representations of complex problems—such as the distribution of antiretroviral medicines— and thereby allow decision making based on evidence and forecasts (Adams 2015; Rottenburg et al. 2015). In this case, indicators were used to forecast different funding scenarios (Figure 2), which were used by the government of Uganda and donors to agree on the funding needed for antiretroviral treatment.

Managing access to antiretroviral therapy as a technical problem immensely increases the speed of broadening access to treatment. Moreover, technologies like indicators help to increase the speed of decision making on various levels—from patient enrolment to donor decisions. The increased speed of broadening access obscures the enormous social acceleration produced by global health projects.

As I will discuss in the next sections, technicisation of mass HIV treatment programmes increases the gap between project time and lifetime. Project time denotes the time given in the form of projects or global health objectives, whereas lifetime refers to lived experiences of spending time with others—be it patients, students, or colleagues. Most notably, lifetime captures the experience of care as time spent with others. The acceleration of time reflects the increasing gap between project time and lifetime experienced as a shortage of time, as Blumenberg suggests. In the case of large-scale HIV treatment programmes, the shortage of time has been particularly drastic in terms of pharmaceuticals. Occupying pharmaceutical supply chain management as a scientific discipline, pharmacists continuously optimise, increase efficiency, and improve the supply of medication to people living with HIV. In this

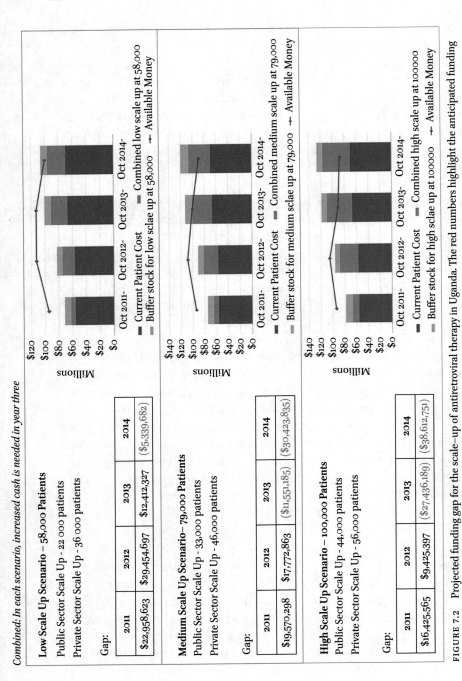

FIGURE 7.2   Projected funding gap for the scale-up of antiretroviral therapy in Uganda. The red numbers highlight the anticipated funding deficit for three different scenarios

SUPPLY CHAIN RATIONALISATION AND ART SCALE-UP SCENARIOS PRESENTATION, 20.1.2012; WITH COURTESY FROM THE TECHNICAL ADVISORY COMMITTEE, MOH technicised world, efficiency is achieved at the expense insight into the pharmacy profession, as I aim to show in the next sections.

3.2   *Extra Time and Regaining Insight*

I encountered the antinomy between efficiency and insight during an evaluation of a TASO project to improve service delivery. The greater the access to treatment at TASO, the more the organisation's infrastructural capacity grew. In 2009, TASO maintained eleven treatment centres around the country, with each site providing around three thousand patients with antiretroviral therapy. In addition, more than ten thousand people were registered as "clients" at each of the treatment centres and routinely prescribed Cotrimoxazole Prophylaxis and offered counselling services to improve adherence to treatment. Each treatment centre saw more than five hundred patients per day on average.

As access to treatment grew, TASO had to expand its physical infrastructure. Waiting areas had to be enlarged to ensure privacy. Outreach services brought HIV-related services into communities but also helped to decongest the treatment centres. Furthermore, the large quantities of medicine dispensed every day required more storage space. The acceleration of treatment access also inevitably meant that TASO employees' workloads increased. Thus, a greater degree of standardisation was required to ensure the efficient "flow of patients" from the initial triage to the final step of receiving antiretrovirals at the dispensary. As they were responsible for the last step at the centres, pharmacists often complained that they were "the last ones to finish".[12]

TASO's growth illuminates the transformations that have occurred as a result of the *projectification* of health care in Uganda, as described by Susan Reynolds Whyte and colleagues (2013). Work at TASO is evaluated on the basis of performance metrics and is subject to what is called implementation research to continuously improve the delivery of health services. These improvements present an unacknowledged challenge to spend time in ways that are productive for global health but often ignore practices that might enrich our understanding of what matters in pharmacy yet cannot really be measured.

Early in the treatment roll-out in Uganda, physicians such as Elly Katabira noted that the rapid scaling up of access to treatment would inevitably increase the demand for qualified health workers. The increase in access also called for dispensing antiretrovirals in order to "be user friendly and [...] deliver[ed] closer to where the action is" (Katabira and Oelrichs 2007, 8), namely at the treatment centres. The pharmacists at TASO's treatment centres took this

---

12   Fieldnotes, October 20, 2010.

approach to heart. Improving pharmaceutical services was of utmost importance, according to Andrew from Medical Access, especially to ensure donor
satisfaction and subsequently a continuous flow of funds. Medical Access initiated a mentoring programme, with the project proposal on "Mentoring for Services Improvement in Pharmaceutical Management at TASO Service Centres"
stating:

> The increase in the number of clients has strained the resources at the
> service centres. This has led to a decline in the level of service quality in
> pharmaceutical management. Sub-optimal performance in pharmaceu-
> tical management is indicated by [...] challenges at some of the service
> centres.[13]

The proposal lists challenges such as "poor stock management," "poor records
management," and, most importantly, "poor medicine counselling". Together,
these challenges carry the risk of leading to "poor treatment compliance"
by patients. The proposal furthermore argues that these challenges require
an innovative intervention. Instead of building capacity through workshops
measured by the number of participants, which only strained the delivery of
health services, the proposal offers a "problem-based approach" to learning.
Mentors would for instance train pharmacists onsite at the treatment centres,
thereby ensuring minimum time away from the pharmacies.

Additionally, the proposal emphasizes an evidence-based approach to
training which would measure the improvement of pharmaceutical practices
according to recent trends in the field of pharmacy (Trap et al. 2010). Randomised clinical trials, originally devised to test medicines, were used in health
service delivery. The model was implemented at four sites, and there were two
control sites. To assess the impact of mentoring, the project measured a range
of key performance indicators. For example, to measure performance in storage practices, the number of "expired medicines" were counted.

During my evaluation of the project, the mentors, who were hired from the
School of Pharmacy of Makerere University, told me that at first the trainees
at TASO expected "someone to come in" and instruct them to "do this or do
that." They also thought that they would be granted a travel allowance, something they were used to from participating in workshops. In contrast, instead
of the mentor telling the trainees "what to do," problem-based learning meant

---

13   Project Proposal *Mentoring for Service Improvement* (2010) provided by Medical Access.

that the trainees had to identify the problems themselves before developing an action plan. It thus "took time for the concept to sink in."[14]

Fred, one of the mentors, recalled how the mentors and trainees identified six to seven problems at each site. The mentors then asked the trainees to identify the most pressing ones that needed to be addressed immediately. Fred's trainees identified their workload at TASO as the main problem. A first measure was to set up a work roster. Initially, the trainees expected that developing a roster, meeting with each other, and assigning different tasks and responsibilities would create even more work. But once the roster was in place, Fred explained, the trainees "saw" the difference and began to "acknowledge" the impact of mentoring. After a few mentoring sessions, the trainees began "to understand," and "you [could] see the things just clicking." As Fred summarised, "the trainees are themselves coming up with improvements."[15]

For Fred, it was crucial that the trainees understood what they were required to do. As Fred pointed out to me, mentoring requires trust. Trainees have to open up instead of fearing disciplinary measures. This trust, in turn, requires empathy with trainees, he maintained. Most importantly, making trainees understand things takes time. As he stated, "I start with small dosages of mentoring and increase the dosages *slowly*" (my emphasis).[16]

Problem-based learning takes extra time but in the long run leads to sustainable change because it is initiated by the trainees themselves, as Fred explained. Instead of disseminating learning materials, mentoring meant listening and understanding everyday work challenges. Fred explained: "I *take extra time* and do more visits in the beginning, otherwise they [trainees] will forget" (my emphasis).[17] These mentoring sessions were very much about sharing work experiences and creating a mutual understanding of what could be improved. As Fred understood the word, mentoring was about empowering trainees to identify problems, develop a solution on their own, and thereby initiate change. However, this empowerment is not possible without making trainees understand that pharmacy is not only about logistics. As he frequently pointed out, pharmacy is more than counting drugs. Pharmacy, he emphasised, is about patient care, which is increasingly sacrificed to the effective management of pharmaceutical logistics.

Sharing experiences and creating insight in terms of Blumenberg's notion of social acceleration were incompatible with the goal of measuring the impact of

---

14 Fieldnotes, October 26, 2010.
15 Fieldnotes, October 31, 2010.
16 Ibid.
17 Ibid.

mentoring through key performance indicators. The project was designed as a randomised clinical trial, undermining the very idea of moving slowly and taking care of each trainee. The mentors I accompanied travelled from Kampala to Jinja, Soroti, Gulu, and Masindi, and then again from Kampala to Masaka, and Rukungiri—essentially to all major regions in the country—within a matter of two weeks. The schedule allowed two days at each treatment centre. The mentoring session started on the first day with the assessment of pharmaceutical practices and key performance indicators. This involved the team of mentors, and myself who administered a fifteen-page form. The data collection required a full day. The second day was dedicated to mentoring, which meant that the team worked together with the trainees to solve individual problems.

The schedule of this project resonates with observations about the thinning out of the complexity of places and social life (Falzon 2009). Most importantly, indicators were hardly able to capture the major aim of mentoring which was to improve trainees' understanding of pharmaceutical logistics, and how this understanding could presumably lead to long-lasting improvements in the broader organisation. At the final presentation to the executive director and the whole board of directors of TASO, we—the team of mentors and myself—tried to account for the qualitative dimensions of learning by providing short case studies alongside the representation of the performance indicators. For the board executive director, however, what mattered most was that the performance indicator to measure storage practices across all TASO sites had recorded a value of 75 percent, meaning that 25 percent of the overall stock comprised expired medicines. The executive director was appalled by the figure and asked: "What went wrong?"[18]

For the board of directors, it was a problem of organisational structure: "Who is responsible for the waste?" They were hardly interested in how the problem had been solved by the trainees. And the response of the executive director was grim: "If the funders see we are wasting resources, we have a problem!" Thus, while trainers like Fred emphasized that learning how to identify problems and develop realistic solutions takes time,—which essentially meant that extra time needed to be spent with the trainees—this effort to create an understanding of pharmacy as a medical practice was unmeasurable.

The mutual exclusion of efficiency and understanding that articulates the experience of the acceleration of time is not limited to organisations such as TASO but pertains to the entire discipline of pharmacy, as I show in the next section. Extra time taken by Fred reflects an attempt to close the gap between

---

18   Fieldnotes, January 4, 2011.

what could be termed *project time* in analogy to Blumenberg's notion of *lifetime*, and world time. Acceleration is an irreversible process that increases the gap between project time and lifetime. This incongruent relationship thrives on efficiency. Taking time to create insight is perhaps better understood as extra time that cannot be measured and accounted for.

3.3 *Regaining Pharmacy*

This experience of acceleration is not limited to TASO and other NGOs that have been mushrooming in Uganda and other sub-Saharan African countries. Consultancies reflect another dimension of this projectification in the domain of academic knowledge production, resulting in a perceived lack of time in which to do professional academic work. Daniel Wight, Josephine Ahikire, and Joy Kwesiga (2014) provide a rare study of the extent of consultancy research in Uganda, revealing that two-thirds of the interviewed researchers had been involved in some kind of consultancy research over a period of three years. According to the authors, a recurrent topic was that consultancies paid better than universities but also took up "a lot of institutional time" (Wight, Ahikire, and Kwesiga 2014, 34). The meaning of "institutional time" remains vague, but I suggest that it hints at the common understanding that academics should spend time in ways that serve the public interest, a notion that conflicts with consultancy work that is based on gaining individual benefit necessary to make a living.

Like many other pharmacists, Andrew was concerned about the enormous income gap between the private and public health sector, which is characteristic of most African countries. He was particularly concerned about the fact that the lack of time because of consulting is equally a lack of time to determine what kind of pharmacy work people should be trained in and what kind of pharmaceutical research should be conducted (see also Mamdani 2011). He referred to the official statistics—which any Ugandan pharmacist would be well acquainted with—that show the ratio of pharmacists to the population in the country was 1:800,000. As the demand for pharmacists increased, various universities around the country, both public and private, started offering pharmacy programmes. Since 1988, the Pharmacy School of Makerere University had produced twenty pharmacists with a BA Pharm degree but had not been able to establish a Masters or PhD programme. New programmes evidently could not procure sufficient teaching staff.

Andrew, who was part of the mentoring project, attributed the problems of pharmacy to the influence of the massive influx of donor money for antiretroviral therapy. Donor-funded projects had increased the demand for pharmacists significantly but mainly for implementing the required standards of pharmaceutical supply chain management. This demand did not necessarily contribute to the improvement of pharmacy as a scientific discipline. It is worth considering a longer passage from Andrew's account:

> I call it [an] insane amount of money. HIV has been in the area for many years, but it's been crazy amounts of money. Like in the beginning, there was the Bush government [as PEPFAR], there was pressure on these institutions to spend money and for them to put pills in a patient's mouth. That was the push. In a way it's been great also for our profession, but in a way, it has also been tricky. Many pharmacists wanted to work for HIV programmes, because of the funding. These programmes are the ones who can pay huge salaries—at the expense of the other areas. But also, now, when you are interviewing people, young pharmacists, they ask you for big salaries when they have zero experience. Without any experience they are asking for this. Now, you may argue that we don't have to live in the past, but sometimes it is unimaginable, you ask someone with zero experience: "What is your expectation?" $2,000, you know! Because this is what their peers working for these [NGO] programmes are earning, that's what they ask for. You know, now you see this everywhere, whether you are working in [the] HIV area or in any other area.[19]

Andrew wondered whether it would still count as pharmacy if the new generation of pharmacists jumped onto the bandwagon of logistics, just for the money, while pharmacists in the public health system were seriously lacking. Other interlocutors critically questioned the reputation of pharmacists in Uganda, asking: "Is pharmacy a profession or a fashion?" and "Do pharmacists still know who they are?"

4   Conclusion

Exploring the acceleration of time in large-scale HIV treatment programmes contributes to the critical ethnography of global health by enriching our understanding of access to treatment as a technical problem and how it shapes new forms of subjectivity, existence, and modes of government. It draws our attention to the technicised worlds of logistics, which is characterised by a growing gap between project time and lifetime and manifested in the experience

---

19   Interview AM, Kampala, Uganda, July 13, 2010.

that time for creating insight is in short supply. Creating insight, as the participants pointed out, is about knowing how things are done, which requires time. Moreover, this experience of acceleration articulates a loss of institutional time because of work devoted to the cultivation of pharmacy as a profession and scientific discipline.

My critical analysis of acceleration is not denying the need for speed, particularly in terms of access to antiretrovirals. In the case of antiretroviral therapy and the necessity to scale up treatment for everyone living with HIV, acceleration is not to be mistaken for technical speed, which is measured in clock time and the number of people who have access to treatment. The acceleration of time suggests that technical efficiency is achieved at the expense of gaining an insight into pharmacy as a profession in the field of global health. This acceleration amplifies the gap between project time and lifetime—that is the actual time given within the limits of projects and the projectified infrastructure of global health as opposed to the time necessary to learn how pharmacy is done.

This gap cannot be closed by doing things faster to increase the efficiency within a set of financial constraints, most notably budget limits. This means that the only option to achieve efficiency is to get cheaper antiretrovirals to save money and continuously cut all unnecessary costs. In fact, scaling up access to treatment means that the number of patients grows, which in turn increases the demand for care and other kinds of treatment. This demand for care requires more pharmacists, and this means greater training and teaching capacities. Most importantly, teaching and learning how pharmacy is done requires time to show, learn, and gain professional experience and an understanding that pharmacy is more than delivering pills to patients, as Andrew pointed out.[20]

Throughout this chapter, I described how the yawning gap between project time and lifetime is produced, how it is reflected upon, and how it is addressed. The notion of extra time reveals how the relationship between efficiency and insight is inverted. For the pharmacists with whom I worked, insight means gaining an understanding of the profession that goes beyond the counting of drugs and, as with any other medical field, seeks to improve care. It means that the pharmacy is less driven by material gains but that it is attuned to the work at hand. Extra time reveals the experience of time for the sake of gaining insight and regaining pharmacy as an existential experience, which is neither measured nor accounted for in the field of global health. In this respect, the

---

20   Interview AM, Kampala, Uganda, July 13, 2010.

technicised lifeworld of pharmacists might be representative of various other medical disciplines that on the surface benefit enormously from the promotion of global health. Acceleration can neither be stopped by reducing speed, nor through more—or less—technical progress. Rather, the critique of the acceleration of time suggests that global health is not only lacking money, resources, or knowledge, but also—crucially—extra time for producing insight.

Bibliography

Adams, Vincanne, ed. 2015. *Metrics: What Counts in Global Health.* Durham, NC: Duke University Press.

Adams, Vincanne, Nancy J. Burke, and Ian Whitmarsh. 2014. "Slow Research: Thoughts for a Movement in Global Health." *Medical Anthropology* 33, no. 3: 179–97. https://dx.doi.org/10.1080/01459740.2013.858335.

Blumenberg, Hans. 1986. *Lebenszeit und Weltzeit.* Frankfurt am Main: Suhrkamp.

Blumenberg, Hans. (1963) 2015. "Lebenswelt und Technisierung unter Aspekten der Phänomenologie." In *Hans Blumenberg: Schriften zur Technik*, edited by Alexander Schmitz and Bernd Stiegler, 163–202. Frankfurt am Main: Suhrkamp.

Espeland, Wendy. 2015. "Narrating Numbers." In *The World of Indicators: The Making of Governmental Knowledge through Quantification*, edited by Richard Rottenburg, Sally E. Merry, Sung-Joon Park, and Johanna Mugler, 56–75. Cambridge: Cambridge University Press.

Falzon, Mark-Anthony. 2009. "Introduction: Multi-Sited Ethnography: Theory, Praxis, and Locality in Contemporary Research." In *Multi-Sited Ethnography: Theory, Praxis and Locality in Contemporary Research*, edited by Mark-Anthony Falzon, 1–24. Burlington, VT: Ashgate.

Fassin, Didier. 2013. "A Case for Critical Ethnography: Rethinking the Early Years of the AIDS Epidemic in South Africa." *Social Science and Medicine* 99: 119–126. https://dx.doi.org/10.1016/j.socscimed.2013.04.034.

Fortun, Mike. 1998. "The Human Genome Project and the Acceleration of Biotechnology." In *Private Science: The Biotechnology Industry and the Rise of Contemporary Molecular Biology*, edited by Arnold Thackray, 182–201. Philadelphia: University of Pennsylvania Press.

Katabira, Elly T., and Robert B. Oelrichs. 2007. "Scaling up Antiretroviral Treatment in Resource-Limited Settings: Successes and Challenges." *AIDS* 21, Suppl. 4: S5–S10. https://dx.doi.org/10.1097/01.aids.0000279701.93932.ef.

Mamdani, Mahmood. 2011. "The Importance of Research in a University: Keynote Address." Paper presented at the Research and Innovations Dissemination Conference, Makererer University, Kampala, Uganda. Accessed April 6, 2021. https://

misr.mak.ac.ug/publication/working-paper-no-3-the-importance-of-research-in-a
-university.

Mugyenyi, Peter. 2009. "Flat-Line Funding for PEPFAR: A Recipe for Chaos." *The Lancet* 374: 292.

Okero, F. Amolo, Esther Aceng, Elizabeth Madraa, Elizabeth Namagala, and Joseph Serutoke. 2003. *Scaling Up Antiretroviral Therapy: Experience in Uganda*. Geneva: WHO (World Health Organization).

Peterson, Kristin. 2014. *Speculative Markets: Drug Circuits and Derivative Life in Nigeria*. Durham, NC: Duke University Press.

Pigg, Stacy Leigh. 2013. "On Sitting and Doing: Ethnography as Action in Global Health." *Social Science & Medicine* 99: 127–34. https://10.1016/j.socscimed.2013.07.018.

Prince, Ruth Jane. 2014. "Introduction: Situating Health and the Public in Africa. Historical and Anthropological Perspectives." In *Making and Unmaking Public Health in Africa: Ethnographic and Historical Perspectives*, edited by Ruth J. Prince and Rebecca Marsland, 1–54. Athens: Ohio University Press.

Redfield, Peter. 2013. *Life in Crisis: The Ethical Journey of Doctors Without Borders*. Berkeley: University of California Press.

Rosa, Hartmut. 2013. *Social Acceleration: A New Theory of Modernity*. New York: Columbia University Press.

Rosa, Hartmut. 2016. *Resonanz: Eine Soziologie der Weltbeziehung*. Frankfurt am Main: Suhrkamp.

Rottenburg, Richard, Sally Engle Merry, Sung-Joon Park, and Johanna Mugler, eds. 2015. *The World of Indicators: The Making of Governmental Knowledge through Quantification*. Cambridge: Cambridge University Press.

Sangaramoorthy, Thurka, and Adia Benton. 2012. "Enumeration, Identity, and Health." *Medical Anthropology* 31, no. 4: 287–91. http://dx.doi.org/10.1080/01459740.2011.638 684.

Stengers, Isabelle. 2017. *Another Science Is Possible: A Manifesto for Slow Science*. Cambridge: Polity Press.

T'Hoen, Ellen, Jonathan Berger, Alexandra Calmy, and Suerie Moon. 2011. "Driving a Decade of Change: HIV/AIDS, Patents and Acccess to Medicines for All." *Journal of the International AIDS Society* 14, no. 1: 1–12. http://dx.doi.org/10.1186/1758-2652-14 -15.

Taylor, E. Michelle, and Ian Harper. 2014. "The Politics and Anti-Politics of the Global Fund Experiment: Understanding Partnership and Bureaucratic Expansion in Uganda." *Medical Anthropology* 33, no. 3: 206–22. http://dx.doi.org/10.1080/014597 40.2013.796941.

Thompson, Edward P. 1967. "Time, Work-Discipline, and Industrial Capitalism." *Past and Present* 38: 56–97.

Trap, Birna, Ebba Holme Hansen, Rete Trap, Abraham Kahsay, Tendayi Simoyi, Martin Olowo Oteba, Valerie Remedios, and Marthe Everard. 2010. "A New Indicator Based Tool for Assessing and Reporting on Good Pharmacy Practice." *Southern Med Review* 3 (4): 4–11.

UNAIDS. 2013. *Global Report: UNAIDS Report on the Global AIDS Epidemic 2013*. Geneva: Joint United Nations Programme on HIV/AIDS (UNAIDS).

Whyte, Susan Reynolds, Michael A. Whyte, Lotte Meinert, and Betty Kyaddondo. 2006. "Treating AIDS: Dilemmas of Unequal Access in Uganda." In *Global Pharmaceuticals: Ethics, Markets, Practices*, edited by Adriana Petryna, Andrew Lakoff, and Arthur Kleinman, 240–71. Durham, NC: Duke University Press.

Whyte, Susan Reynolds, Michael A. Whyte, Lotte Meinert, and Jenipher Twebaze. 2013. "Therapeutic Clientship: Belonging in Uganda's Mosaic of AIDS Projects." In *When People Come First: Critical Studies in Global Health*, edited by João Biehl and Adriana Petryna, 140–65. Princeton, NJ: Princeton University Press.

WHO. 2003. *Scaling Up Antiretroviral Therapy in Resource-Limited Settings: Treatment Guidelines for a Public Health Approach*. Geneva: WHO.

Wight, Daniel, Josephine Ahikire, and Joy C. Kwesiga. 2014. "Consultancy Research as a Barrier to Strengthening Social Science Research Capacity in Uganda." *Social Science & Medicine* 116: 32–40. http://dx.doi.org/10.1016/j.socscimed.2014.06.002.

Index

Printed in the United States
by Baker & Taylor Publisher Services